NATURAL CATEGORIES AND HUMAN KINDS

The notion of "natural kinds" has been central to contemporary discussions of metaphysics and philosophy of science. Although explicitly articulated by nineteenth-century philosophers like Mill, Whewell, and Venn, it has a much older history dating back to Plato and Aristotle. In recent years, essentialism has been the dominant account of natural kinds among philosophers, but the essentialist view has encountered resistance, especially among naturalist metaphysicians and philosophers of science. Informed by detailed examination of classification in the natural and social sciences, this book argues against essentialism and for a naturalist account of natural kinds. By looking at case studies drawn from diverse scientific disciplines, from fluid mechanics to virology and polymer science to psychiatry, the author argues that natural kinds are nodes in causal networks. On the basis of this account, he maintains that there can be natural kinds in the social sciences as well as the natural sciences.

MUHAMMAD ALI KHALIDI is Associate Professor of Philosophy at York University, Toronto.

NATURAL CATEGORIES AND HUMAN KINDS

Classification in the Natural and Social Sciences

MUHAMMAD ALI KHALIDI

CAMBRIDGE
UNIVERSITY PRESS

CAMBRIDGE
UNIVERSITY PRESS

University Printing House, Cambridge CB2 8BS, United Kingdom

Cambridge University Press is part of the University of Cambridge.

It furthers the University's mission by disseminating knowledge in the pursuit of
education, learning and research at the highest international levels of excellence.

www.cambridge.org
Information on this title: www.cambridge.org/9781107521728

© Muhammad Ali Khalidi 2013

First published 2013
First paperback edition 2015

A catalogue record for this publication is available from the British Library

Library of Congress Cataloguing in Publication data
Khalidi, Muhammad Ali, Professor.
Natural categories and human kinds : classification in the natural and
social sciences / Muhammad Ali Khalidi.
pages   cm
Includes bibliographical references and index.
ISBN 978-1-107-01274-5 (Hardback)
1. Categories (Philosophy)   2. Classification.   I. Title.
BD331.K43 2013
001.01'2–dc23
2012044074

ISBN 978-1-107-01274-5 Hardback
ISBN 978-1-107-52172-8 Paperback

For Diane,
one of a kind,
and Layla,
in a category by herself

Contents

Figures

Preface

As an undergraduate majoring in physics in Beirut, Lebanon, I once came across a smartly illustrated volume by Philip Morrison and Phylis Morrison, entitled *Powers of Ten*. Intriguingly subtitled, "A Book About the Relative Size of Things in the Universe and the Effect of Adding Another Zero," it was a photographic journey through 42 orders of magnitude, from the scale that corresponds to the size of the observable universe (10^{25} m) to the scale of elementary particles (10^{-16} m). There was a familiar scene depicted somewhere in between these two extremes, on a scale of the order of 1 m, of a woman and a man picnicking in a park in downtown Chicago. The book zoomed out from the picnic to the city, continent, planet, solar system, galaxy, and beyond, and then zoomed in to the cells in the man's hand, to the molecules and atoms constituting them, and eventually to the quarks inside the protons and neutrons in the nuclei of the atoms. It was bracing to experience the universe as a series of logarithmic steps from the inconceivably large to the unimaginably small. This picture of the world, which is the one conveyed to us by modern science, suggests realms of existence arranged in levels, from smallest to largest. But these are not self-contained, compartmentalized levels like the floors in an apartment building, since there are intricate relations and interactions between the levels, or *domains*, as I shall call them later in this book. Additionally, the domains are not discretely arranged in a hierarchy. Much of the universe is a jumble of domains, some coexisting at the same spatiotemporal scale and within the same regions of space-time and others overlapping partially, or, to use a term that I have used elsewhere, "crosscutting" each other. Modern science has evolved an array of disciplines, subdisciplines, and interdisciplinary research programs to study this complex multiplicity, each with its toolkit of categories, generalizations, and methods. This book is about the assortment of categories that scientists have devised to study the multifaceted nature of reality, and specifically which of these categories are valid or, to use the philosophical jargon, correspond to 'natural kinds'.

Many philosophers favor a picture, which may be as old as Aristotle, in which there is a relatively small set of privileged categories, and according to which each individual object in the universe belongs properly to one category, which conveys its essence. Essentialism, which may have a bad name in the culture at large, is alive and well in academic philosophy departments, though many of its proponents would deny that the philosophical doctrine corresponds to the set of popular ideas that bear the same name. Be that as it may, I will argue that the central claims of philosophical essentialism have either not been adequately justified or are at variance with what modern science tells us. Philosophical doctrines should not find themselves out of step with the scientific worldview – at least that is what a naturalist stance in philosophy would recommend. Some other philosophers, and many academics outside of philosophy departments, tend to think that the unfeasibility of essentialism is glaringly obvious. They may then go on to add that it is equally clear that all our categories, whether scientific or folk, are creative inventions, constructed by human beings to fulfill various practical and social purposes, but without any serious claim to drawing an accurate picture of the universe. To think otherwise is to be guilty of a kind of anthropocentric hubris. This social constructionist (or conventionalist) position is often pitted against the essentialist position in a dialogue of the deaf. What I aim to do in this book is to defend an alternative position that is neither essentialist nor social constructionist (or conventionalist). It is a naturalist position, which takes into account the discoveries of various scientific disciplines while at the same time trying to derive general conclusions about the validity of our categories.

The pigeonholes into which we slot objects in the world are convenient devices that enable us to fulfill our explanatory needs and predict future contingencies, but insofar as they succeed in this regard, they do so precisely because they are attuned to regularities and patterns in the natural world (including the social world). I will argue that there is no conflict between the claim that our categories serve our purposes and the claim that these categories correspond to natural kinds, provided that they serve genuinely epistemic purposes. Our classification schemes and taxonomic practices enable us to focus on some features of reality while neglecting others in order to make sense of these patterns of constancy and change.

The title of this book is a bit perverse. Many readers might instead expect *Natural Kinds and Human Categories*. That is because *natural kinds*, the types or sorts that the natural world is divided into, are usually contrasted with *human categories*, which human beings concoct to serve their idiosyncratic interests. Kinds of natural objects are also sometimes contrasted with the categories into which we divide ourselves and our conspecifics. So 'human categories' can refer either to those categories devised by humans or to the categories into which human beings are divided. But this book questions some of the assumptions inherent in these distinctions. First, as I have already suggested, I will argue that there is often a close connection between the kinds that are present in the world and the categories that we invent to understand the world, and second, I will defend the position that some of the types into which humans are divided can also be considered natural kinds.

In Chapter 1, I take on two philosophical theses about natural kinds that have prevailed in the philosophical literature during the past few decades: metaphysical Realism and essentialism. Metaphysical Realism holds that natural kinds are a type of universal; that is, that they are abstract entities over and above their members. This Realist (as opposed to Nominalist) position considers natural kinds to be more than just collections of particulars. Though this position may be justified by certain philosophical considerations, it is of limited use if our aim is to identify which kinds are natural. That is because it does not give us a way of distinguishing natural from nonnatural kinds. It simply says that the natural ones correspond, metaphysically speaking, to universals rather than sets of particulars. This view is often coupled with essentialism about natural kinds, which continues to be the dominant theory of natural kinds among contemporary analytic philosophers. Unlike metaphysical Realism, essentialism purports to put forward criteria for distinguishing natural from nonnatural kinds. On an essentialist view of natural kinds, each natural kind is associated with a set of properties that are necessary and sufficient for membership in the kind, modally necessary (i.e., pertain to the kind or to its members in every possible world), intrinsic, microphysical, and discoverable by science. But I argue that the essentialist view of natural kinds is difficult to maintain in the face of modern science and argue that each of these conditions except the last is either inadequately supported or out of step with our current knowledge of the natural world.

Chapter 2 introduces my own positive account of natural kinds, according to which natural kinds are epistemic kinds, which I develop by situating it in relation to the views of Locke, Mill, Quine, Dupré, and Boyd. I find something to agree with in the views of each of these philosophers, though I also take issue with each of them in some way. Natural kinds correspond to those categories that enable us to gain knowledge about reality. Since science is the enterprise dedicated to acquiring knowledge about the world, natural kinds are identified by the various branches of science. Of course, we do not know which categories will remain part of our settled scientific account of the world, so any endorsement of the current categories of science is corrigible and subject to revision in light of future inquiry. This view is defended against the charge that it is too restrictive as well as the charge that it is too liberal. The charge that the account is too restrictive concerns the existence of natural kinds outside of scientific inquiry, corresponding to the folk categories of ordinary language, but I argue that many folk categories are not introduced to serve an epistemic purpose and should not therefore be taken to provide an accurate account of the kinds that exist in reality. As for the charge that the account is too liberal, it amounts to identifying certain further conditions that natural kinds must satisfy (i.e., in addition to being discoverable by science). Some of the most prominent of these conditions have already been examined and dismissed in Chapter 1 in the course of criticizing essentialism. In this chapter, I discuss other conditions, which are also found problematic: that natural kinds must be discrete or have sharp boundaries, that natural kinds cannot crosscut one another but must be arranged in a nonoverlapping hierarchy, and that each natural kind must be associated with a causal mechanism that maintains its associated properties in a state of equilibrium, i.e., "homeostatic property clusters" (Boyd 1989, 1991). While Boyd's account is too restrictive in that it posits a causal mechanism that keeps all the properties in the cluster in homeostasis, it does point to the importance of grounding the epistemic efficacy of natural kinds in causal relations. Building on recent philosophical work, I therefore propose a "simple causal theory" of natural kinds (Craver 2009). Hence, the epistemic conception of natural kinds leads naturally to a metaphysical account in terms of causality.

Chapter 3 defends the view that natural kinds can occur in the 'special sciences' just as much as in the basic sciences. There is a widespread assumption that the kinds of the special sciences are importantly different from those of the basic sciences. The former are often thought to be functional kinds, which are either just multiply realizable disjunctions of "lower-level" kinds or else reducible to them. Moreover, special-science kinds and properties are thought not to have causal efficacy since all the causal work must be done at a "lower level." It is also sometimes argued that there are no laws in the special sciences or, if there are, they are very different from the laws of the nonspecial sciences. I argue against each of these claims, while focusing on a particular natural kind from fluid mechanics, *Newtonian fluid*, and a closely associated property, *viscosity*. These arguments provide further support for the "simple causal theory" of natural kinds introduced in Chapter 2. But that theory is challenged by the claim that natural kinds will be too numerous and ineffectual to be worth the name. I defend the account against this objection and provide further evidence for the idea that systems of natural kinds can crosscut one another because they pertain to different aspects of the natural world. This leads me to introduce the notion of a scientific *domain*, which I distinguish from the more widespread idea of "levels" of reality.

The claim that natural kinds are epistemic kinds implies that categories derived from the biological and social sciences can also be natural kinds. In defense of this claim, in Chapter 4, I critically examine several attempts to distinguish kinds in the natural sciences from those in the biological and social sciences. Some philosophers think that biological and social kinds cannot be natural kinds for the very reason that special-science kinds generally cannot. But others hold that they cannot for other reasons, the most prominent of which are explored in this chapter, and I argue in each case that they give us no grounds for thinking that biological and social kinds cannot be natural kinds. I consider the view that biological kinds are etiological kinds, individuated by causal history rather than causal powers. I also examine the distinction between "eternal kinds" and "copied kinds" (Millikan 1999), the latter being kinds whose members resemble each other not as a matter of natural law but as a result of a copying process. Then I counter the view that social kinds are conventional (Searle 1995); though the most conventional of kinds are not natural kinds, it is clear that many social kinds are not conventional, or not entirely so. Hacking (1999, 2002) claims that human kinds can be interactive whereas natural kinds cannot, but some natural kinds also come into existence as a result of human intervention and they can interact in various ways with our thoughts and actions. Finally, Griffiths (2004) holds that at least some social kinds are normative or evaluative in character, a feature that distinguishes them from kinds in the natural sciences. However, normativity is by no means a feature of all social kinds, and when it is, it can be detected. I conclude that categories in the biological and social sciences are not fundamentally different from those in the natural sciences and that biological and social kinds can be natural kinds as well.

Chapter 5 looks at several case studies drawn from a range of sciences in order to test the claims about natural kinds that I have made so far. In the spirit of philosophical naturalism, I examine a number of widely accepted and controversial kinds to ascertain whether they can be considered natural kinds. The case studies are drawn from basic physics and chemistry (*lithium*); chemistry, materials science, and polymer science (*polymer*);

biochemistry, physiology, and virology (*virus*); physiology, medicine, and oncology (*cancer* and *cancer cell*); and psychiatry and cognitive science (*attentiondeficit/ hyperactivity disorder* [ADHD]). These case studies enable me to corroborate and amplify some of the claims that I make in earlier chapters and also to further elaborate and illustrate these claims. Though in all cases, I conclude that the kinds examined are good candidates for natural kinds, I also encounter some kinds along the way that I argue are probably not natural kinds.

Finally, in Chapter 6, I attempt to show that this naturalist approach to natural kinds is compatible with realism about kinds. Though I do not engage in a full-blown defense of scientific realism (not to be confused with metaphysical Realism, discussed in Chapter 1), I give some reason for adopting a realist attitude towards natural kinds. In doing so, I further clarify the relationship between natural kinds and properties and the role of causality in the proper characterization of natural kinds. In defending a realist account of natural kinds, I counter the charge that natural kinds are determined by our interests or perspective on the world. Though my account of natural kinds is pluralist and does not set an upper limit on the number of natural kinds that may exist, it holds that these kinds really exist in the world. It is common for philosophers to express realism about kinds in terms of the claim that kinds are human- or mind-independent, but I reject this way of grounding realism since it threatens to rule out all psychological and social kinds. More importantly, to be real, a kind need not be independent of human beings or their minds; it must simply be manifested in the world (a world that includes the human mind). The surest way to ensure that our categories identify real kinds is to pursue a scientific method that serves epistemic purposes. Finally, I relate this discussion to the "social constructionist" position about categories or kinds; though some versions of the social constructionist thesis are compatible with my natural-ist position, other social constructionist claims are either trivial or false.

******

This book has taken me a few years to write but I have spent many more years thinking about some of the questions that I address in these pages. A number of people have helped me think through these issues, often setting me straight on certain points, indicating the deficiencies of my arguments, or revealing certain lines of argument that had not occurred to me. It is difficult to recall all the conversations that I have had over this period and I am sure I am forgetting to credit some of them, but I would like especially to acknowledge the help and encouragement of Ian

Hacking, John Heil, Tom Nickles, and Stephen Stich. I would also like to thank Bana Bashour and Hans Muller of the American University of Beirut for inviting me to present some of this work to a conference in Beirut in May 2011. There, I was very fortunate to receive feedback on an earlier draft of Chapter 2 from both of them as well as from all the participants in the conference. My wonderfully supportive colleagues at York University have also heard me present some of this material and have been excellent philosophical interlocutors. In addition, I would like to acknowledge the support of two travel grants from the Social Science and Humanities Research Council of Canada (SSHRC) and a grant from the International Conference Travel Fund of the Faculty of Liberal Arts and Professional Studies at York University, which enabled me to present portions of this work at three conferences, the Metaphysics of Science conference (Nottingham, 2009), the Philosophy of Science Association Biennial Meeting (Montreal, 2010), and the Society for Logic, Methodology, and Philosophy of Science in Spain (Santiago de Compostela, 2012). Thanks are due to audience members and participants at all three conferences for comments and discussion.

When it comes to more specific debts concerning this book, a number of people have taken the time to help me with feedback and advice. First, I would like to thank two anonymous referees for Cambridge University Press who provided astute comments, constructive criticism, and vital encouragement. Others who gave me very useful comments are my students, Rami Elali, who read Chapter 1, and Abigail Klassen, who read the entire manuscript; I have had stimulating discussions about natural kinds and social kinds with both of them. I am also grateful to Abigail Klassen for helping to prepare the index for this book, thanks in part to a Minor Research Grant from the Faculty of Liberal Arts and Professional Studies at York University.

Students in two graduate seminars at York University, one on natural kinds and another on the philosophy of social science, helped me by letting me try out some of my half-baked ideas. I am also grateful to a graduate student, Orsolya Csaszar, for valuable research assistance on ADHD, and to an undergraduate student, Rachelle Innocent, for writing a research paper on this topic, which helped to acquaint me with the literature on ADHD and helped me work on section 5.6. Parts of Chapter 1 of this book, specifically section 1.7, have appeared in a previously published paper, "The Pitfalls of Microphysical Realism," *Philosophy of Science* 78 (2011), 1156–1164. I am grateful to the journal for permission to reprint portions of that paper.

At Cambridge University Press, I would like to express my immense gratitude to Hilary Gaskin for encouraging me to undertake this project and for overseeing its early and later stages. Thanks are also due to Anna Lowe for patience and diligence, especially as two deadlines passed for delivering the manuscript. My great appreciation also goes to Joseph Garver for expert copy-editing, and to Thomas O'Reilly for help and advice during the production stages.

Members of my far-flung extended family have been remarkably indulgent of my foibles and shortcomings while I worked on this project. They know who they are and how much they mean to me. Finally, I have dedicated this book to my wife Diane and my daughter Layla because that is the only way I know to thank them. I did not include my son Zayd in the dedication, not because I know of some other way to thank him but because I dedicated an earlier book to him. To paraphrase Oscar Wilde, to have one book dedicated to you may be regarded as a misfortune, but to have two looks like carelessness.

CHAPTER I

Metaphysical Realism and essentialism about kinds

I.I KINDS OF THINGS

We are a classifying species. We recognize not just individuals but *kinds* of things, and we sort individuals into kinds. Among the myriad kinds we identify are *protons* and *antineutrinos*, *lithium* and *roentgenium*, *polystyrene* and *DNA*, *radioactive decay* and *polymerization*, *stars* and *meteorites*, *Newtonian fluids* and *gases*, *viruses* and *cancer cells*, *homologies* and *larvae*, *child abuse* and *Alzheimer's disease*, *hysteria* and *ADHD*, and *permanent residents* and *refugees*. These include kinds of entity or object, process or state, and so on. In the face of such a proliferation of kinds, philosophers are prone to ask whether all of them are on a par, or whether some are real and others merely ersatz, artificial, or nominal. Some philosophers would regard only a small minority of such groupings as real or natural. They would claim that the natural kinds are a tiny subset of the kinds that we have identified in the course of our everyday activities and in the course of scientific theorizing about the world. On this way of seeing things, not all categories identified in our natural language, nor even all those featured in scientific discourse, ought to be taken to pick out real kinds of things. Many, if not most, are simply convenient groupings, with limited utility for some purpose or another, but without a claim to "carving nature at its joints." This supposed contrast between categories that really correspond to the divisions in nature and those that are merely useful crutches designed to enable us to get by in the world (let alone those that are entirely artificial and fail to serve any practical purpose) is the focus of this chapter. I intend to examine the various criteria and desiderata that have been put forward to distinguish natural from artificial kinds, and will try to determine which of them, if any, should be taken as a mark of the natural.

Consider any set of individuals endowed with various properties, whether human beings, artifacts, terrestrial organisms, clouds, celestial bodies, samples of chemical substances, or elementary particles. Each individual in this set will typically have a large number of properties, and any attempt to systematically describe the whole collection will inevitably involve sorting individuals into groups. Now imagine that a human observer, call her Eve, surveys this scene and wonders how she is to make sense of these individuals, each with its own physical dimensions, spatial location, trajectory, causal powers, patterns of behavior, and so on. After a period of close observation, Eve hits upon a system for dividing the individuals into groups, which helps her make sense of it all, which has explanatory power, and on the basis of which she is able to make surprising predictions. Her sorting scheme consists of a system of categories, $K_1, \ldots, K_n$, each including a number of individuals among its members, based on the properties possessed by those individuals. Each of these categories is associated in her language with a general term; each such general term picks out a particular *kind* of individual. If she finds herself in a philosophical, rather than a purely scientific, mood, she may mull over a number of questions. Having sorted these individuals into a system of kinds, she might ask herself the following: Are these the kinds to which these individuals *really* belong? Do divisions between the various kinds correspond to the world's own divisions, or are they merely a reflection of my perspective? Moreover, can they be further split into subkinds, or further lumped into superkinds? Is there a single unique way of sorting them into kinds, or are there a number of different ways of doing so? If there is no unique way of doing so, are some systems of kinds privileged over others, or are they all on a par?

Having formulated these questions and considered them, Eve might raise a further question: How are we to tell which of these categories really correspond to the world's own divisions? Is there some way of doing so beyond our usual ways of discerning which categories succeed and advance our knowledge and which do not? It is not as if some categories come with a further proof of authenticity or a seal of approval that informs us that they are genuine while the others are not. Thus, Eve may conclude, the question concerning which kinds are real (or natural) would seem to reduce to one about which categories figure in our best theories of the world, or form part of our settled knowledge of nature. It is not that our best theories and settled knowledge actually *determine* which kinds exist, but rather that they serve as the best *guide* to the existence of the kinds of things in the world. We have no other way of delineating genuine groupings from bogus ones, we can imagine Eve concluding. Ultimately, Eve's conclusion is the one that I will be arguing for in this book. But in this chapter I will first examine other proposals for establishing which kinds are natural and which are not. Various ways of distinguishing the real categories or 'natural kinds' have been proposed, and philosophers have advanced several answers to the question, what makes a kind natural? Some of these have explicitly been put forward as accounts of natural kinds, but others are either implicit in such answers, or emerge in slightly different contexts to distinguish real from unreal kinds of entities.

Before proceeding, there are a few preliminary issues to be clarified. One such issue concerns philosophical methodology. How should we go about adjudicating the issue of what constitutes a natural kind? One traditional philosophical approach would recommend analyzing the concept *natural kind*, but this immediately raises the question of what we are to go on when we perform such an analysis. Some philosophers might posit a direct metaphysical intuition that would enable us to identify the criteria by which to distinguish natural from nonnatural kinds. But this seems to assume that we have an intuitive knack for discovering the underlying nature of reality, which is not an assumption I am prepared to accept without further justification. Moreover, we cannot go on common parlance and attempt to explicate our common usage of the expression, since 'natural kind' is a philosophical term of art, first introduced into discussions by John Venn (1876), following John Stuart Mill (1843/1974), who used the expressions "real kind" and "true kind."[1] And merely analyzing the usage of these philosophers would be a historical exercise. Instead, it would be more fruitful to adopt the methodology of "reflective equilibrium" (Goodman 1954/1979), throwing into the hamper various relevant factors. One such factor concerns our convictions as to the categories generally regarded as paradigmatic natural kinds, such things as elementary particles, chemical elements, chemical compounds, biological species, and perhaps a few others (beyond that, things are more controversial). A philosophical account of natural kinds that deems all or many of these to be natural kinds is to be preferred over one that does not, other things being equal.[2] We should also factor past philosophical usage into the equation; it would count against a view of natural kinds if it does not cohere at all with at least some previous philosophical discussions of natural kinds (and this is where the

[1] Although Hacking (1991, 110) credits Venn with coining the expression, Venn (1889, 83) credits Mill, saying that "he introduced the technical term of 'natural kinds' to express such classes as these." But Mill tends to use the terms 'real kind' and 'true kind' instead of 'natural kind'; I will discuss Mill's view in Chapter 2.

[2] Some contemporary essentialist accounts of natural kinds have the consequence that biological species are not natural kinds (Ellis 2001; Wilkerson 1993). While such accounts should not be dismissed out of hand, this consequence can be considered a drawback.

views of Mill and others would at least be relevant). An account that did not overlap at all with previous ones may well be accused of changing the subject. A third factor that should figure in our deliberations concerning natural kinds consists of a set of considerations drawn from actual scientific practice as to which categories are regarded as genuine as opposed to mere artifacts, and as to the methods that are used to make such judgments. The attempt to take scientific evidence seriously in this philosophical inquiry is in line with a "naturalist stance" in contemporary philosophical discussions. Moreover, scientific evidence can also be brought to bear in a different way in this philosophical inquiry. If a philosophical account of natural kinds holds that all natural kinds should have some feature F, and if our current best scientific theories of what are commonly regarded as natural kinds tell us that these kinds lack F, then that would cast doubt on this philosophical proposal. (Of course, it may be possible for us to save F at the expense of deeming that those kinds that lack F are not natural kinds after all, but that is a price we should try to avoid paying, other things being equal.) Yet another factor to subject to reflective equilibrium is the sum of considerations derived from other areas of philosophy, such as discussions of natural laws, properties, and causation, as well as broader questions in epistemology and philosophy of language. In the final analysis, there will be choices to be made – for instance, in regarding how to rank these considerations, and when to revise convictions in one area at the expense of others. I will endeavor, whenever there are judgment calls to be made, to make them explicit and to flag them as such.

Another issue worth pausing to consider is a terminological one. The term 'natural kind' has come to be central to this philosophical debate and I have used it several times in the previous paragraphs. As I have already indicated, the term has a venerable history and there is a clear rationale for using it, since it points to a contrast between categories that exist *in nature* and those that do not (existing perhaps only in our minds). But the term is also unfortunate, since it may suggest a connection with the natural sciences (conventionally, physics, chemistry, and biology) as opposed to the social sciences. Now, some philosophers would indeed restrict natural kinds to the natural sciences (and some would further restrict them to a subset of those sciences and to a subset of the categories therein, as we shall see), but the very use of the term should not lead us to prejudge the issue. At least, I want to consider it an open question and will try to determine whether the restriction of natural kinds to the natural sciences is justifiable. The word "natural" in the term 'natural kind' is more plausibly regarded as alluding to the fact that the kinds in question are really found in nature or in the universe (not merely in the mind or in language). It might have been better to use Mill's expression "real kind" instead, but unfortunately that expression has never caught on and is not a widely used expression. Since 'natural kind' has come to be used to distinguish real from nonreal kinds, that is the term I will deploy. Another issue raised by the use of the term 'natural kind' concerns the appropriate complementary term. Of course, the least controversial expression to denote the complement of 'natural kind' is 'nonnatural kind', though that is not a commonly used term and is not very informative. On the other hand, some of the terms that have been used in this connection seem committed to substantive answers to questions that, once again, I would like to keep open. 'Nominal kind' suggests that kinds that are not natural exist in name alone, or are present only in language. 'Artificial kind' implies that they are a product of human artifice. 'Artifactual kind' conjures up human-made artifacts. Thus, for lack of a better alternative, I will opt for the more neutral 'nonnatural kind', despite its awkwardness. Furthermore, as I will use it, the term 'kind' on its own is meant to encompass both natural and nonnatural kinds. I will also use the term 'category' to denote a kind-concept, a concept that refers to a kind, whether natural or not. Roughly speaking, a 'category' belongs to our language, theories, or discourse whereas a 'kind' pertains to the world. Finally, I will tend to italicize the names of kinds and categories when they are being considered as kinds or categories but not when discussing their instances or manifestations (though the distinction is occasionally hard to draw).

1.2 KINDS AND UNIVERSALS

Some philosophers would say that what distinguishes natural from non-natural kinds is that the former correspond to real entities, and that these entities are abstract objects endowed with metaphysical reality. This is Realism in the classical sense, as found in various guises in the history of philosophy, from Plato to David Armstrong. In what follows, I will use 'Realism' (uppercase *R*) when referring to the thesis that properties and kinds refer to universals, distinct metaphysical entities, rather than sets of particulars. This thesis is not to be confused with a more limited thesis of realism (lowercase *r*) about kinds, which regards them as objective features of reality (to be discussed in Chapter 6), not necessarily corresponding to distinct metaphysical entities like universals.

Kinds, like properties, are thought on this Realist view to have metaphysical reality over and above the particulars that belong to those kinds.

Why posit an entity, such as a kind or property, as distinct from the members of that kind or the instances of that property? Historically, philosophers have put forward several considerations for doing so, but two will suffice for our purposes. One is that the very same collection of individuals can sometimes constitute all and only the members of more than one distinct natural kind. If we were to identify a kind with its individual members, then we would sometimes be unable to maintain that these were indeed distinct kinds. The kind *creature with a heart* is often said to be actually coextensive with the kind *creature with a kidney*, yet they seem to be distinct kinds. More plausibly, in the phylogenetic taxonomy of living organisms a genus sometimes contains a single species (or a family a single genus, and so on). Even though the individual members of the species are identical to the members of the genus, the species and genus would seem to be distinct natural kinds. A second reason for positing properties over and above their instances, or kinds over and above their members, and for thinking of them as entities in their own right, is that we often have occasion to refer to them or quantify over them in our theoretical or scientific pursuits. Armstrong (1980/1997, 106) uses state-ments such as the following to make this point:

(1) There are undiscovered fundamental physical properties.
(2) Some zoological species are cross-fertile.

In these statements, it is not a trivial matter to paraphrase away occurrences of the terms 'properties' and 'species', or to replace the statements with ones that refer only to sets of particulars. Hence, we seem to be committed to the existence of properties and kinds in some of the statements that we make. This argument is particularly effective against metaphysical anti-Realists, or Nominalists, since many of them follow Quine's ontological dictum that "to be is to be the value of a variable."[3] If we find ourselves quantifying over properties and kinds and if we are unable to do away with them in our considered scientific theories, then we need to posit entities that correspond to them, or to admit such entities into our ontology. What sort of entities, then, would correspond to properties and kinds?

One historically influential view of properties is that they are identical with universals, which can be construed either as being transcendent (along the lines of Plato's 'forms') or immanent in the particulars that possess the relevant properties. On the latter view, which has been

---

[3] Curiously, (2) is mentioned by Quine (1948/1953), though he does not explain how a nominalist might rephrase it.

defended by Armstrong (1978a, 1978b, 1989), universals are wholly present in each of their instances, as nonspatiotemporal parts of them. For example, the universal *positive charge of 1.6×10^{-19} C* is present in each proton particle, though it is not a detachable part of each such particle. This view of universals has certain unintuitive consequences, since it entails, for example, that something can be entirely present in two distinct instances at the same time. Moreover, it countenances such things as 'parts' that are neither spatiotemporal nor detachable. Should one reject this entire conception of universals based on the fact that these entities violate some of our most basic intuitive assumptions about reality? Lewis (1983, 345), who is not exactly sympathetic to this view, thinks not. After all, he asserts, our intuitions about such matters "were made for particulars". Be that as it may, positing strange entities of this sort exacts a price.

According to Realism about properties, each real property corresponds to a universal, a metaphysically independent entity that is repeated in each of its members. In Armstrong's terminology, each universal is the "truth-maker" for a particular having a certain property. How would Realism deal with kinds? Kinds differ from properties in that their instances are individual entities or objects (as well as, perhaps, events, processes, and so on), while properties are instantiated by property instances, which are sometimes referred to by metaphysicians as 'tropes' or 'modes'. The kind *elephant* has individual elephants as its instances (e.g., Dumbo), while the property *gray* has particular manifestations of shades of gray as its instances (e.g., Dumbo's grayness). In addition, kinds are "associated with"[4] collections of properties (the nature of this association will be discussed shortly, as well as in section 6.2), since individuals belong to kinds on the basis of possessing a number of properties, and indeed there may be nothing more to being a member of a kind than possessing a certain set of properties. Incidentally, I will assume that members of kinds are similar to each other because they share at least some of these properties. Some philosophers (e.g., Heil 2003) think that there can also be brute similarity between individuals and property instances, and that membership in a kind is based on similarity. But I find this notion of brute similarity to be obscure and prefer to understand similarity in terms of shared properties; in this,

---

[4] I am following many contemporary authors in using this (somewhat vague) locution. One exception is E. J. Lowe, who thinks that kinds are characterized by properties just as their instances are. He thinks that it is acceptable to say that "certain kinds are characterizable by certain characterizing universals," and that this is consistent with "saying that particular instances of those kinds are *also* characterizable by those universals" (Lowe 2004, 155; original emphasis). But the kind *elephant* is not gray in the same way as Dumbo is gray.

I agree with Mill (1843/1974, IV vii §4), who writes: "And this resemblance [among members of a kind] itself is not, like resemblance between simple sensations, an ultimate fact, unsusceptible of analysis. Even the inferior degree of resemblance is created by the possession of common characters." I will also assume, following a number of contemporary philosophers, that properties are individuated by their causal powers and are closely associated with them (Armstrong 1978b). Properties are sometimes considered either *categorical* or *dispositional*, the latter being causal powers that are manifested under certain conditions. But this distinction does not seem to run very deep, and some philosophers have proposed that every property has both a dispositional and categorical aspect (Heil 2003). Further, properties can be determinable (e.g., *mass*) or determinate (e.g., *mass of 67 kg*), but we shall see in due course that the properties associated with natural kinds tend to be determinate rather than determinable.

What does Realism about properties have to say about kinds? There would seem to be two ways of accommodating kinds on the Realist picture. On one account, in addition to the fact that each property corresponds to a universal, the kind associated with a collection of properties also corresponds to a universal. Consider the kind *proton*, whose members are individual protons. This kind is associated with the properties of having a *positive charge of 1.6×10^{-19} C*, having a *mass of 1.7×10^{-27} kg*, having *spin ½*, and so on. Each of these properties corresponds to a universal, so the question arises as to the relationship between the universal corresponding to the kind *proton* and the universals corresponding to each of its associated properties. Possessing a certain collection of properties is both necessary and sufficient for being a proton, so we would expect the universal that corresponds to the kind *proton* to have some intimate connection to the universals corresponding to each of its associated properties. How exactly this is to be spelled out is a delicate matter.

Many proponents of universals already recognize that they need to posit (higher-order) relations between universals, specifically relations corresponding to natural laws. Indeed, some, like Armstrong (1992/1997, 164–165), take this to be one of the attractions of admitting universals into our ontology – namely, that universals are involved in providing "ontological correlates" to true statements of natural laws. On Armstrong's view, the truth-makers for laws of nature are the necessary connections that obtain between some universals. Thus, if it is a law of nature that negative charges and positive charges attract each other, this is made true by a necessary connection between the property-universal *negative charge* and the property-universal *positive charge*. Now if the kind *proton* corresponds to a universal in its own right, we would need to explain the relationship between the universal corresponding to the kind and the universals corresponding to each of its associated properties. We would also need a way of distinguishing this relationship from that obtaining between universals linked by natural laws, as well as from that obtaining between universals structurally linked to other universals (e.g., the universal *proton* and the universal *hydrogen atom*, or in the other direction, the universal *proton* and the universal *up quark*). There may indeed be ways of spelling out the truth-makers for these relationships between universals, but this does not seem to have been worked out in detail, and there are considerable problems that confront some attempts to do so.[5]

This brings us to the second way of dealing with kinds on a Realist view of properties: One might hold that the kind itself does not correspond to a single metaphysical entity, but rather to a conjunction of such entities. The Realist thesis would then apply not to the kind so much as to the cluster of properties that members of the kind have in common. In some of his work, Armstrong (1997, 67) casts doubt on the need for separate universals to correspond to natural kinds, writing that "it is not clear that we require an independent and irreducible category of universals to accommodate the kinds." But if that is the case, it does not seem as though we have endowed the kind itself with any metaphysical status, but rather have done so for its associated properties. The kind would then correspond to a conjunction of universals rather than a single universal. Moreover, conjunctions of universals do not seem to have any more claim on reality than conjunctions of particulars (e.g., David Armstrong and Louis Armstrong, or my favorite pen and the Rock of Gibraltar). In an inventory of the objects that exist in the universe, we would not count individuals twice over, once on their own and again as members of two-somes or couples. This would seem to apply to the realm of universals too. Moreover, though Armstrong (1978b, 30–39, 1989, 84) thinks that conjunctions of universals are themselves universals (unlike, say, disjunctions or negations of universals), other Realists dissent from this judgment. For instance, Ellis (2001, 89–90) thinks that there are no conjunctive universals corresponding to conjunctive properties. Though he does admit conjunctions of properties that correspond to natural kinds, he does not appear to justify this exception to the denial of conjunctive universals.

---

[5] For some of the problems faced by structural universals, see Lewis (1986); for a response, see Armstrong (1986).

Even if we satisfactorily resolve the question as to whether a conjunction of universals can be considered a universal, when it comes to kinds, identifying them with conjunctions of universals is particularly unhelpful. If one were to consider the kind *proton* to correspond to the conjunction of properties associated with that kind, this would not constitute a ringing endorsement of the existence of kinds. We would have no more reason to think that the kind *proton* exists as a conjunction of the properties associated with protons than we have to believe in the existence of the conjunction of any two or more of those properties. If the kind *proton* is taken to be equivalent to a conjunction of property-universals, it would have no more claim to existence than the conjunction of *positive charge of 1.6×10^{-19} C* and *spin ½*. But that conjunction is not a natural kind whereas *proton* is. The difference between the two cannot be explained by a view that considers kinds to be simply conjunctions of universals. Realists could say that the natural kinds are only those that correspond to single properties, not those kinds that are associated with a number of different properties. But if they were to say that the only natural kinds are those that correspond to single properties, they would be left with a rather unsatisfactory account of natural kinds. It would turn out that having a *positive charge of 1.6×10^{-19} C* is a natural kind, and that its members include all protons as well as all pion particles and others, but that *proton* itself is not a natural kind. However, an account of natural kinds that deems natural only those kinds that correspond to single properties is not really an account of natural kinds at all, since few of the natural kinds that are widely accepted are thought to coincide with single properties.[6] One might as well say that there are no kinds over and above the properties with which they are associated.

To summarize, the first version of Realism owes us some account of the relationship (presumably, a necessary connection) between the universal that corresponds to the kind and those corresponding to its associated properties. And the second version seems not to endow kinds with an independent metaphysical existence. However, there is another problem with considering kinds to be universals. A more pertinent problem, at least for our purposes, is that it does not give us a way of distinguishing natural from nonnatural kinds. To see this, consider again Realism about

---

[6] Another problem with this move for Realists is that many of them take properties and kinds to belong to different ontological categories (e.g., Ellis 2001; Lowe 2006). Hence, the universal corresponding to the kind cannot be the same as the universal corresponding to the property even in the case of a single-property kind.

properties. Armstrong's Realist view of properties purports to tell us what it is to be a property – namely, that it is to correspond to a universal or cluster of universals – but it does not tell us which properties there are. In other words, it may be an account of what it is for something to be a property, but it does not equip us with a method of distinguishing real or natural properties from nonproperties. If Armstrong's account of properties is extended to kinds, identified either with universals or conjunctions of universals, the same issue recurs. Armstrong explains what universals are, but when it comes to saying which universals exist, he says that should be left up to the sciences to specify. He writes:

in the present age, we take ourselves to have advanced beyond the epistemic state of nature, and to have sciences that we speak of as 'mature.' There we will find the predicates that constitute our most educated guess about what are the true properties and relations. (1992/1997, 166)

Someone might object to this by saying that our scientific theories, which are based on our best epistemic practices, are being held to determine what exists. However, that objection would surely be misconceived. It is not that Armstrong believes that the content of our ultimate scientific theories will determine which properties there are in the sense of *endowing* them with that status. Rather, these theories will merely determine which properties there are in the sense of *ascertaining* what is most fundamental to the universe. This should not be surprising if one deems that science is (or eventually will be) successful in uncovering the ultimate constituents of the universe, or which types of things make up the world. But it goes to show that this account does not purport to tell us how to differentiate natural from nonnatural kinds (no more than it does so for properties).

The Realist view of kinds, which considers kinds (like properties) to be universals endowed with metaphysical reality, has problems endowing kinds with an independent metaphysical status. These problems may have solutions but I will not have anything more to say about them. It seems clear that kinds should not just be identified with their actual members, since there could be coextensive kinds that are not identical, yet it is not clear how exactly to characterize them, metaphysically speaking. Even though this is a worthwhile question, the main issue that I am trying to address is not one about the metaphysical status of kinds, but rather one about the distinction between natural and nonnatural kinds. Armstrong may be right to think that this is a question for scientists to address in every particular case, but this leaves open a number of important questions as to whether all scientific categories correspond to natural kinds or whether only some do, which ones do, whether they pertain only to the basic sciences or whether they can occur in the special sciences, and so on.

1.3 KINDS AND ESSENCES

Metaphysical Realism about natural kinds is often conjoined with essentialism about kinds, even though there is no necessary connection between the two theses. They are often held together because they provide answers to complementary questions (roughly, *what* are natural kinds, and *which* kinds are natural?). Unlike metaphysical Realism, essentialism attempts to provide a criterion for distinguishing natural from nonnatural kinds; it holds that the former have essences (or are associated with essential properties) while the latter do not. An essentialist account of natural kinds tends to specify certain criteria that serve to delineate the class of essential properties, and hence that of natural kinds. These criteria can be used to assess any candidate for natural kindhood, to see whether it is indeed a natural kind. A kind K, on this picture, is associated with a number of properties, $P_1, \ldots, P_n$, and these properties are distinguished by certain recognizable features, though not all essentialists agree on these features. In recent philosophical discussions, the features that are most commonly associated with essentialism about natural kinds would seem to be the following:

(a) *Necessity and sufficiency*: Each of the properties associated with a natural kind is possessed by every individual that belongs to that kind, and any individual possessing all of them belongs to the kind in question (cf. Ellis 2001, 22; Soames 2002, 15).

(b) *Modal necessity*: (i) Natural kinds are such that they are associated with the same set of properties in every possible world; (ii) the properties associated with a kind are such that an individual member of the kind would possess them in every possible world in which that individual exists (not just in the actual world) (cf. Ellis 2001, 21; Wilkerson 1988, 35).

(c) *Intrinsicality*: The properties associated with a natural kind are possessed by an individual member of that kind independently of that individual's relations to anything else in the universe (cf. Ellis 2001, 20; Wilkerson 1988, 29).

(d) *Microstructure*: The properties associated with a natural kind are "underlying" microphysical properties rather than macrolevel properties (cf. Wilkerson 1988, 41).

(e) *Discoverability by science*: The properties associated with a kind can be ascertained by scientific inquiry and are those properties that will eventually feature in a completed science (cf. Wilkerson 1988, 29).

Of these features, (a) and (b) (i) have a different status from the rest, since they pertain to the relation between the kind and its associated properties, whereas all the rest concern aspects of the associated properties themselves. If K is the kind in question and $P_1, \ldots, P_n$ are its associated properties, (a) and (b) (i) say something about the relation between K and the Ps. Meanwhile, (b) (ii), (c), (d), and (e) say something about the Ps themselves; they tell us what kinds of properties they are supposed to be. Though all essentialists may not hold all of the above theses, they all hold at least some of them. Since I will be arguing that all but the last thesis are problematic, I will take it that this casts doubt on essentialism. In the rest of this chapter, I will be considering each of these aspects of essentialism in turn, starting in this section with the last one, discoverability by science.

Essentialists think of natural kinds as corresponding to real divisions in nature. These kinds are not just conventionally distinguished by human inquirers; they would be discerned by any rational observer undertaking an objective examination of the universe. What sets natural kinds apart is that each is associated with a real essence, a property or set of properties that is unique to it and is the basis for differentiating it from other natural kinds. These properties are ones that human inquirers are generally capable of ascertaining and science is the enterprise that is dedicated to doing so. For essentialists, the properties associated with each kind, or a subset of them, are the essence of that kind. Crucially, the essential properties point to the occurrence of yet other properties, and science is able to project these properties from one instance to others. Consider the chemical element *lithium* (atomic number 3). Lithium is a silver-colored, soft metal, with low density, which is highly corrosive, is flammable, is a good conductor of heat and electricity, reacts with water, and has a melting point of 180.5 °C. Moreover, lithium is an ingredient of lithium-ion batteries, which have numerous technological applications. It has also been used in pharmacological contexts as a mood-stabilizing drug due to its neurophysiological effects on the human brain. The essence of this natural kind, as with all other members of the periodic table, is held to be its atomic number. The other, more superficial, properties of lithium flow from this essential property. It should be emphasized that the essential property is often discovered after a large number of other, more superficial properties have already been observed. But science aims to discover the essential properties and uses them to explain the more superficial properties. Essentialists sometimes also distinguish these superficial properties from accidental properties, which are not part of the essence at all. For example, an individual sample of lithium may weigh 100 g or it might be in the shape of a rectangular prism, but these properties are not part of the essence of the kind to which it belongs.

Essentialism is surely right to draw attention to the fact that discoverability by science is a key feature of natural kinds, and that science is primarily interested in kinds that enable us to project from one or a few samples to many others, or from past samples to future ones. What distinguishes natural kinds is that they are associated with properties that point to yet other properties in a reliable way, allowing us to make strong inductive inferences. Projectibility is therefore a central feature of natural kinds and is perhaps the main reason that scientists attempt to look for them.[7] This much is common ground among essentialists and nonessentialists. Indeed, nonessentialist philosophers also maintain that natural kinds are discoverable by science and constitute the basis for projection and inductive inference. This view of natural kinds can be traced back at least to the nineteenth century. For instance, Venn (1876, 49) observes that the collection of statistical information in science is carried out against an assumption that natural kinds exist: "Such regularity as we trace in nature is owing, much more than is often suspected, to the arrangement of things in natural kinds, each of them containing a large number of individuals." More recently, Kornblith (1993, 36) writes that the existence of natural kinds, and the fact that scientists can reliably infer the presence of some properties from the presence of others, is the key to understanding and explaining the world. In other words, it is the clustering of properties in natural kinds that makes the world objectively knowable. As Chakravartty (2007, 170) puts it: "Properties, or property instances, are not the sorts of things that come randomly distributed across space-time. They are systematically 'sociable' in various ways." He uses the metaphor of sociability to indicate that some properties are regularly co-instantiated with others, and natural kinds are, as it were, the locus of this sociability. Natural kinds enable us to use the instantiation of some properties to infer the presence of others. Thus, there is widespread agreement among essentialist and nonessentialist philosophers alike that natural kinds are

---

[7] Usually, predicates or terms are said to be projectible (Goodman 1954), meaning that they are capable of figuring in strong inductive inferences. I will often also say that properties or kinds are projectible, meaning that the predicates or terms that denote them are.

the grounds for rich inductive inferences, and that science is concerned to discover natural kinds. But essentialists do not just hold that natural kinds play an important inductive role in scientific inquiry. They tend to impose further conditions on natural kinds and their associated properties, as I indicated above, and I will argue in the following sections that these additional conditions are not warranted.

1.4 DEFINABILITY

Necessary and sufficient conditions are much maligned nowadays, and for good reason. Hence, the claim that natural kinds are linked to their associated properties in the manner of necessary and sufficient conditions may seem to be prone to the usual objections against such conditions. The standard philosophical objection against definitions consisting of necessary and sufficient conditions is that such definitions may be modified in the course of belief revision, without revising the concepts or categories that they allegedly define. Especially in the context of scientific inquiry, it often transpires that the necessary and sufficient conditions that were once thought to define a certain concept are found not to hold as we learn more about a certain subject matter. Yet, on many accounts of scientific concepts or the meanings of scientific terms, such discoveries do not always result in a change of meaning. Hence, it is concluded, necessary and sufficient conditions cannot supply the contents of concepts, particularly scientific concepts. But this objection simply does not arise in this particular context. The link that is thought to obtain between a kind and its associated properties is crucially different from the link that supposedly holds between the *concept* of a kind and its associated definition. Though the concept of a kind should reflect the kind itself in a situation of omniscience, it will not do so in general. At the end of inquiry, if our theories correctly identify the kinds that exist in reality and if we emerge with a true account of their associated properties, then the kind-concept or category will be defined in terms of statements or propositions describing the properties associated with that kind. But then the objection cannot be raised, since there will be no room for belief revision. Therefore, the standard objection to necessary and sufficient conditions does not apply to the claim that a natural kind is connected to a set of properties by relations of necessity and sufficiency, since this relationship is supposed to be revealed at the end of inquiry.

Although there are other objections to necessary and sufficient conditions, many of them apply to such conditions as they pertain to kind-concepts rather than to kinds themselves. What, then, is wrong with associating a natural kind with a requisite set of properties, in such a way that possession of those properties is both necessary and sufficient for membership in that kind? I will argue not that this state of affairs *cannot* obtain – it surely can – but rather that it may not be the only way in which a natural kind might be linked to a set of associated properties. The standard alternative to kinds associated with necessary and sufficient conditions ('definable kinds', or to use a term that some scientists employ, *monothetic kinds*) is what is sometimes referred to as "cluster kinds" or (*polythetic kinds*). Though cluster theories are familiar to philosophers and psychologists alike, they are mostly understood as theories of concepts or categories, rather than theories of kinds. In what follows, I will investigate some of the features associated with cluster theories considered as theories of the relationship between kinds and their associated properties. My aim will be to ascertain whether there are any convincing reasons for restricting natural kinds to monothetic kinds, kinds that are associated with a set of properties, all of which are both necessary and sufficient for kind membership.

To avoid confusion I will try to be a bit more precise concerning the difference between monothetic and polythetic kinds. A monothetic kind is one that is associated with a property or set of properties each of which is singly necessary for membership in the kind and all of which are jointly sufficient. A polythetic kind is one that does not satisfy this condition. Members of a monothetic kind possess all and only the same properties, *qua* members of that kind, whereas members of a polythetic kind may not possess all and only the same properties *qua* members of that kind. In particular, if a kind is associated with a complex construction of properties, such as $K_1 = P_1 \& (P_2 \vee P_3)$, or $K_2 = P_1 \& (P_2 \vee P_3) \& P_4$, then we cannot consider such a kind monothetic, on pain of stripping the distinction of any significance. The whole point of a cluster kind is that there is no unique set of properties that all and only members of that kind possess by virtue of being members of that kind. Two individuals can be members of K_1 not by virtue of possessing exactly the same set of properties; for example, individual i_1 might possess just P_1 and P_2, while i_2 possesses just P_1 and P_3. Hence, K_1 and K_2 are not characterized by necessary and sufficient conditions for membership as ordinarily understood. If necessary and sufficient conditions were watered down in such a way as to allow these kinds, then the distinction between monothetic and polythetic kinds would disappear.

There is a subset of polythetic kinds that deserves some special attention. The examples of kinds that I gave above both involve disjunctive constructions, but they cannot be considered *purely* disjunctive kinds. In other words, the disjunctive terms do not allow us to divide the kind into disjoint sets associated with distinct properties. Of course, it is a trivial truth of logic that any disjunction of the form, say, $P_1 \vee (P_2 \& P_3)$ can be rewritten as $(P_1 \vee P_2) \& (P_1 \vee P_3)$, thus perhaps masking its outward disjunctive structure. Still, the disjunctive nature of the kind in question can be demonstrated by showing that there can be two members of the kind that share no properties in common. If we call the kind associated with this cluster of properties K_3, then i_1 can be a member of K_3 by virtue of possessing property P_1, while i_2 can be a member by virtue of possessing $P_2 \& P_3$. Hence, there can be two members of this kind that possess *no* properties in common. This can be taken to be the mark of a disjunctive kind: a kind that can have two or more members that do not possess any properties in common.

Having laid out the distinction between monothetic and polythetic kinds (and distinguished a subset of polythetic kinds – namely, disjunctive kinds), we can go on to ask whether polythetic kinds can be natural kinds. What grounds might essentialists give for claiming that only monothetic kinds can be natural kinds? Let us start by considering the subset of polythetic kinds that I have called pure disjunctive kinds. There are three reasons for doubting the existence of purely disjunctive kinds. First, Armstrong (1989, 82–83) articulates a reason why disjunctive *properties* do not exist. He starts from the plausible premise that properties are closely related to causal powers. The properties an individual possesses are identical with the causal powers that individual has. If two properties are associated with the very same causal powers, then those two properties are identical. Now consider an individual that possesses a property P_1 but not P_2. This individual also possesses the disjunctive property $P_1 \vee P_2$. But possessing that property obviously does not bestow any further causal powers on the individual. Since the disjunctive property does not endow the individual with any causal powers not endowed by P_1, it cannot be truly distinct from P_1. Plausible as this argument sounds, it might not be straightforwardly transferable to kinds, since it is not obvious that the same intimate connection obtains between *kinds* and causal powers, which would be the crucial premise in the corresponding argument. But there is a second argument against purely disjunctive natural kinds such as $K_3 = P_1 \vee (P_2 \& P_3)$. If K_3 were considered a natural kind, then there could be two members of K_3, i_1 and i_2, where i_1 possesses property P_1, i_2 possesses properties P_2 and P_3, and i_1 and i_2 possess no other properties in common. In this case, there would seem to be no basis for considering them to belong to the same *kind*. But if i_1 and i_2 are not members of a kind, then K_3 cannot be a kind, much less a natural kind. Though this argument leans heavily on a pre-philosophical notion of what it is for two things to be of the same kind, it provides us with some reason for precluding disjunctive kinds from being natural kinds. The third and most compelling consideration for doubting the existence of disjunctive kinds has to do with projectibility. In the previous section, I mentioned that projectibility is the most widely agreed upon characteristic of natural kinds, and may in fact be the very reason for positing natural kinds in the first place. It seems clear that disjunctive kinds like K_3 are *not projectible*. If individuals i_1 and i_2 both belong to K_3, then we cannot project from the fact that i_1 possesses property P_1 that i_2 will also possess P_1, since i_2 may belong to K_3 on the basis of possessing properties P_2 & P_3. This is the most powerful consideration against allowing disjunctive kinds to be natural kinds, and unless there is some compelling reason for thinking otherwise, I will take it that natural kinds cannot be structured as *pure* disjunctions.[8]

Having doubted the existence of purely disjunctive kinds, what should we say about other polythetic kinds, which are not disjunctions but may have structures that embed disjunctions, such as K_1 and K_2 above? In these cases, any two members of the kind will always share some properties, and so the above considerations against disjunctive kinds do not apply. As far as projectibility is concerned, these kinds may be more weakly projectible than monothetic kinds, but they can be projectible nonetheless. If membership in kind K_4 is based on possessing at least seven of ten properties $P_1, \ldots, P_{10}$, then if i_1 and i_2 both belong to K_4, we can make a probabilistic inference from the fact that i_1 possesses a certain property to the fact that i_2 does also. That is, we will still be able to project from some members of the kind to others in the case of a polythetic kind, though our projections will be probable rather than certain. A trickier case involves a polythetic kind in which any two members share at least one property but no single property is shared by all members. Imagine a kind K_5 associated with a set of three properties, P_1, P_2, and P_3, such that the possession of no single property is necessary but the possession of any two is necessary and sufficient for membership in the kind. Then it could happen that

---

[8] Some philosophers inspired by Wittgenstein, e.g., Bambrough (1961), countenance categories with a disjunctive structure, along the lines of "family resemblance" concepts. But even though some folk or common-sense categories may be structured in this way (e.g., *game*), I would venture that these categories do not correspond to natural kinds, precisely because they do not seem to be projectible.

some individual i_1 belongs to the kind on the grounds that it possesses P_1 and P_2, i_2 on the grounds that it possesses P_2 and P_3, and i_3 on the grounds that it possesses P_3 and P_1. Such a kind may also be dismissed as nonnatural on the basis of a consideration similar to one of the considerations that were used to disqualify K_3. The specific reason in this case is that the three individuals, i_1, i_2, and i_3, do not have any single property in common; whatever property any two of them have in common is not shared by the third. So they cannot be considered members of the same kind. Nevertheless, there is an important difference between K_3 and K_5 – namely, that even though there is no single property that all members share, for any two members there will be some property that they have in common. This preserves projectibility of a very weak sort.

Given that polythetic kinds that are not purely disjunctive can still be projectible, though some are more strongly projectible than others, what reason might there be for restricting natural kinds to monothetic kinds? Two reasons might be put forward for placing such a restriction on natural kinds. One reason for thinking that polythetic kinds are not natural, whereas monothetic kinds are, might be that the conditions for membership in the former are clear and precise, while conditions for membership in the latter are vague or imprecise. Since natural kinds correspond to real entities, there should be a determinate answer in each case whether some individual i is a member of kind K. If there is no such determinate answer, it may be argued, then we do not have kinds with well-defined boundaries, and that may give us some grounds for doubting that such kinds are real. But, contrary to appearances, monothetic kinds are not superior to polythetic kinds in this respect. If membership in a polythetic kind is determined by possession of a cluster of properties, that does not mean that the conditions for membership are in some cases undefined, or even vague. The relationship between a polythetic kind and its associated properties can be specified in such a way that there is a fully determinate answer in each case whether an individual i belongs to a kind K. Indeed, even when membership in a kind is determined by having a cluster of properties that is weighted in a certain way, there may yet be a definite answer as to whether a particular individual belongs to that kind. Weighted cluster theories are sometimes construed as providing an account of "graded membership" in a kind, so that higher-scoring individuals are "better" or more typical members than lower-scoring ones, and those that fall just short of the threshold score are considered borderline members. But weighted cluster theories can just as well be taken to provide a definite cutoff between members and nonmembers. Thus, the difference between the two kinds of kind is not that monothetic kinds have well-defined boundaries whereas polythetic kinds do not. Rather, the main difference between polythetic kinds and monothetic kinds is that the latter furnish us with a *simple* criterion to dismiss certain kinds as nonnatural. If possession of a set of necessary and sufficient properties is taken as a formal condition on a kind to be natural, then that gives us a convenient test for weeding out nonnatural kinds. The advantage of associating natural kinds with necessary and sufficient conditions is that it gives us a surefire way of dismissing some kinds as nonnatural. But the fact that this condition provides a clear criterion for distinguishing natural kinds is surely not grounds enough for thinking that it marks a distinction in reality between natural and nonnatural kinds. To think so is to assume that reality conforms to what is convenient for us.[9]

One might go further on this score. Let us suppose that some polythetic kinds are such that there are no strict conditions on membership, and let us say that how many or which properties need to be possessed by an individual is not precisely determined. Would this jeopardize the claim of such polythetic kinds to be natural kinds? It is not obvious that it would (as I will argue in section 2.4). A kind with vague boundaries may yet be real for all that. Some essentialists hold that kinds with vague boundaries are not natural, but their grounds for thinking so are unclear.[10]

There might still be a lingering suspicion that nature does not cluster properties together in a loose way, and that the real kinds in the world should be united by a set of properties that are shared by all and only members of that kind. But what grounds are there for thinking that natural kinds must conform to such a pattern and not to a clustering pattern? A second reason that might be offered is that the most basic kinds found in fundamental physics and chemistry conform to the pattern of monothetic kinds (e.g., all and only protons have the following properties: *positive charge of 1.6×10^{-19} C, mass of 1.7×10^{-27} kg, and spin ½*). But the fact that some archetypal examples of natural kinds are structured in this way does not mean that all natural kinds should be. And if these kinds are held to be the *only* natural kinds, this begs the question in favor of the microstructural kinds of fundamental physics and chemistry, which is precisely one of the questions that is up for discussion, and will be given further

---

[9] I should emphasize that I do not take essentialists to consider this to be a sufficient condition on natural kinds. There are surely all manner of non-natural kinds that can be specified in terms of necessary and sufficient conditions. At best, this would provide a necessary condition on natural kindhood, and that is surely how most essentialists would view it.

[10] See Ellis (2001, 19–20), but note that other essentialists, e.g., essentialists about biological species, are not opposed to vague boundaries (LaPorte 2004, 22).

consideration shortly (see section 1.7). Moreover, I will argue later (section 5.2) that even chemical elements are not generally monothetic. For the moment, it is worth noting that polythetic kinds are also found in some of the natural sciences, particularly in biology.[11] Indeed, biological species, long considered the archetypal natural kinds, have come to be perceived by many philosophers and biologists to be polythetic kinds, since it is now well established that there is no set of necessary and sufficient phenotypical or genetic properties that all and only members of a species share (on the latter point, see, e.g., Hull 1965; Sober 1980). A proponent of necessary and sufficient conditions might insist at this point that biological species are not natural kinds at all, as some essentialists have recently done (Ellis 2001; Wilkerson 1993).[12] But if the reason given for doing so is that their members do not possess a set of necessary and sufficient properties for membership, then that case does not seem to have been made. In short, essentialists have not put forward a general argument for the conclusion that all natural kinds ought to be monothetic kinds, characterizable by a set of necessary and sufficient conditions. In the absence of this requirement, it does not appear that there are other formal or logical conditions that can be placed on natural kinds aside from the restriction on purely disjunctive kinds.

1.5 MODAL NECESSITY

Perhaps the most distinctive feature of essential properties, and the one that is most widely associated with them, is that they are supposed to be necessary to their bearers. One way of expressing this claim is to say that an essential property is possessed by its bearer in every possible world in which the bearer exists. Essential properties are not merely contingent properties, which hold of their bearers in the actual world but not in all other possible worlds. What does essentialism say about kinds? There would seem to be two ways of extending the essentialist thesis to natural kinds. If essential properties are possessed by their bearers in every possible world, then perhaps essential kinds are ones whose members belong to them in every

[11] With biological species in mind, Russell (1948, 443) objected to Keynes' account of natural kinds precisely on the grounds that kinds need not be monothetic: "If a 'natural kind' is defined by means of a number of properties $A_1, A_2, \ldots, A_n$ (not known to be interdependent), we may, for some purposes, consider that an individual which has all these qualities except one is still to be counted a member of the kind."

[12] Other biological essentialists have embraced monothetic essences that advert to historical origin rather than phenotypical or genetic properties (see section 1.6).

possible world, and if all natural kinds are essential kinds, then this applies to them too. This is the first way of extending the necessity thesis to natural kinds: Natural kinds are such that their members belong to them in every world in which those members exist. If i is a member of natural kind K in the actual world, then it is a member of that kind in every world in which i exists (and hence i also possesses all the properties associated with K; this is thesis (b) (ii) in section 1.3).

To avoid any misunderstanding, it is worth mentioning that necessary membership in a kind is compatible with there being no set of necessary and sufficient properties associated with that kind, in the sense of the previous section. Consider again the cluster kind K_4, associated with a set of ten properties, $P_1, \ldots, P_{10}$, where possessing any seven of these properties is sufficient for membership in the kind (but no specific property or set of properties is necessary). Suppose now that an individual possessing any of these properties possesses them in every possible world in which that individual exists. These properties would then be necessary to their bearers, and any individual possessing any seven of them would do so necessarily, and hence necessarily belong to the kind, even though there is no set of properties that is necessary and sufficient for membership in the kind. It could be that i_1 necessarily belongs to K on the basis of possessing $P_1, P_2, \ldots, P_7$, while i_2 necessarily belongs to K_4 on the basis of possessing $P_2, P_3, \ldots, P_8$. Thus, the absence of a necessary condition on an individual's belonging to a kind is no obstacle to an individual *necessarily* being a member of that kind. Conversely, there might be a set of necessary and sufficient properties associated with membership in a kind, yet individual members of that kind might not belong to it necessarily – that is, in every possible world in which those individuals exist. A stock philosophical example may help to illustrate the point. Suppose that there is a set of properties necessary and sufficient for belonging to the kind *bachelor* (say, being human, adult, male, and unmarried). It does not follow that a member of the kind *bachelor* is necessarily a bachelor – that is, would be so in every possible world in which that individual exists.

Although the distinction is not always made clearly by proponents of essentialism, there is another way of extending the necessity thesis to natural kinds. According to this second thesis of modal necessity (this is thesis (b) (i) in section 1.3), a natural kind is one that is necessarily associated with a certain set of properties. That is, if a natural kind K is associated with properties $P_1, \ldots, P_n$, in the actual world, then K is associated with those very same properties in every possible world in which the kind is instantiated. This second thesis says that a natural kind is one that is associated with all the same properties in every possible world. (Again, this has nothing to do with being associated with a necessary condition for membership, since cluster kinds could be associated with the same cluster of properties in every possible world.)

Does the first necessity thesis follow from the second necessity thesis, or vice versa, or are they logically independent? Let us start with the second thesis. If a kind K is associated with the same set of properties in every possible world, it does not follow that individual members of the kind belong to that kind in every possible world in which those individuals exist. To illustrate, if the kind *proton* is associated in every possible world with the property of having a *positive unit charge* (among other properties), it does not follow directly that all actual individual protons would be protons in every possible world in which they exist. The protons that exist in this world may not be protons in other possible worlds; at least nothing in the first thesis implies that they would. So the second thesis does not imply the first. What about the converse? If all members of K are members of that kind in every possible world in which those individuals exist, does it follow that K is associated with the same set of properties in every possible world? Consider the kind *proton* again and let us suppose that each individual proton is a proton in every possible world in which it exists. Does it follow that the kind *proton* is associated with the same set of properties in every possible world? Unless that were the case, the claim that each individual proton is necessarily a proton is undetermined. What would it mean to say that every proton is necessarily a proton, though what it is to be a proton may vary from world to world? Although the first thesis may not imply the second, unless the second thesis is true, the first rings rather hollow. It appears that the first *presupposes* the second, in the sense that if the second thesis is false, the truth value of the first is undetermined. Be that as it may, I will examine each thesis on its merits to see whether either represents a reasonable demand on natural kinds.

Start with the first thesis, which says that members of a natural kind belong to that kind necessarily. What grounds might there be for thinking that all natural kinds should be such that their actual members belong to them in every possible world in which those members exist? It may be said that there is a correlation between those kinds that we deem paradigmatically natural and those that pertain to an individual across possible worlds (and conversely between those deemed unnatural and those that merely apply in this world or in some possible worlds but not others). For example, according to this view, I am human and would be so in every possible world in which I exist; I am also a father but presumably would not be so in some other possible worlds in which I exist. Perhaps grounds could be given for thinking that the connection holds generally. Kinds like *human* that are real or genuine surely pertain to an individual no matter what, whereas kinds like *father*, which are allegedly not real or nonnatural, are ones that apply to an individual merely contingently.

There are a couple of problems with this line of thought. First, claims of necessity regarding individuals, or *de re* necessity, are notoriously difficult to justify and it is not clear how to go about adjudicating them. How an individual actually is, what that individual's properties are, and what kinds it belongs to, are matters that we can inquire about and we have means of resolving. But how that individual would be in another possible world, or indeed across all possible worlds, is not something that we have ready methods for determining. Scientific inquiry, which contemporary essentialists tend to think will supply us with the true natural kinds (though some would add that not *all* scientific kinds are natural), does not seem to be in the business of making such claims or in a position to justify them. Scientific claims tend not to concern individuals and when they do, they tend not to support this modal claim. Consider the essentialist claim that a particular proton could not have been anything but a proton in every possible world in which it exists. Science seems largely silent on how protons would be in other possible worlds, so it is unclear how one would either confirm or deny this essentialist claim on strictly scientific grounds. What about scientific counterfactuals that refer to what might have been the case? A quick glance at these types of statements in science does not provide clear support for essentialism, since the counterfactuals entertained by scientists often revoke purportedly essential properties.[13] Essentialists may respond by saying that "counter-essential" statements in science need to be reinterpreted in some way, but it is not obvious that there will always be a reinterpretation that would do justice to the original statement. To make matters worse, sometimes there is *indirect* evidence from scientific theories that casts doubt on the modal necessities posited by essentialists. There are many interactions involving protons in which they are transformed into other kinds of particles.[14]

---

[13] To cite two such recent "counter-essential" statements by contemporary physicists, Hawking and Mlodinow (2010, 160) write: "If protons were 0.2% heavier, they would decay into neutrons, destabilizing atoms," and Greene (2011, 64) states: "If the electron mass in another bubble universe were a few times larger than it is here, electrons and protons would tend to merge, forming neutrons and thus preventing the widespread production of hydrogen."

[14] Similar points can be made concerning other elementary particle interactions. For instance, in beta-minus decay, a neutron is transformed into a proton along with an electron and an

When iridium nuclei are bombarded with protons, antiprotons are produced as a result. In this interaction, as it is typically described, protons are transformed into antiprotons. Hence, an individual proton may not remain a proton in the actual world. Moreover, there is a widely accepted modal principle that says that if an individual can come to belong to a different kind in the actual world, then that same individual could have belonged to that kind in another possible world. From the fact that a proton *can become* an antiproton in the actual world, we might reasonably conclude that that proton *could have been* an antiproton in some other possible world. Thus, it seems that essentialists are wrong to insist that a proton could not have been anything but a proton in every possible world in which it exists. In response, essentialists may reject the inference from temporal possibility to counterfactual possibility. But that principle is well established and seems innocuous, if any modal principle is (Khalidi 2008). Alternatively, essentialists might insist that when we create an antiproton by bombarding an iridium nucleus with a proton, we do not thereby transform a proton into an antiproton. Rather, we annihilate a proton and create an antiproton in its stead. This is importantly different from transforming the very same particle from a proton into an antiproton, which is what essentialists would deny is possible. However, physics does not sanction this interpretation of the interaction and there is no basis in science for supporting it. Moreover, even if it could be supported in this case, there are other cases in which that interpretation is unavailable; for example, the posited but unobserved process of proton decay, whereby protons decay into antiprotons. This is clearly different from a process of proton annihilation, such as one whereby a proton encounters an antiproton and both are annihilated. Finally, there is another problem with essentialist statements about elementary particles in particular. According to current physical theory, elementary particles are *in principle* indistinguishable from one another in the actual world. Since we cannot "label" or "tag" an individual proton, any claim about the properties or whereabouts of an individual proton in this world is impossible in principle to verify. Hence, a claim of *de re* necessity concerning a particular proton is arguably vacuous.

For at least some of the best candidates for natural kinds, either there is no scientific support for the claim that individual members of those kinds belong to them necessarily, or there is indirect evidence against it.

---

antineutrino. This pertains to leptons as well as hadrons, since, e.g., a muon can decay into an electron, neutrino, and antineutrino.

This holds not just for kinds of elementary particles but for other paradigmatic natural kinds as well. An atom of one chemical element can be transformed into an atom of a different chemical element by a process of radioactive decay. Furthermore, on at least some accounts of biological taxonomy, an individual member of a species S_I can become a member of a different species S_2 as a result of a speciation event in which one population branches off from S_I, thereby resulting in two new species, S_2 and S_3, and rendering S_I extinct.[15]

The second, and arguably more fundamental, problem with this modal thesis is that it is not clear what grounds there are for holding that there is a general connection between a kind being real or natural and its applying to its members necessarily. The reality of the kind itself is not obviously linked to the way in which the kind applies to its individual members across possible worlds. In response to this point, it might be objected that there is indeed a way of drawing a connection between the naturalness of kinds and their modal status. Perhaps part of what it means to say that a kind is natural is that it would be the way it is not just in the actual world, but in every possible world. But the most plausible interpretation of that claim would render it equivalent to the second modal thesis, which says that a natural kind is one that is associated with the very same set of properties in every possible world. This thesis might gain some plausibility from the view that the real kinds in nature are the ones that are associated with certain properties necessarily; that is to say, in every possible world, not just in this world. This may be the intuitive reasoning that lies behind the thesis that natural kinds are necessarily the way they are. Can this line of reasoning be made more persuasive and be grounded in something more substantive than a bare intuition?

The second modal thesis about natural kinds has been most closely associated with the work of Saul Kripke (1980) and Hilary Putnam (1973, 1975), who have argued that natural kind terms, such as 'water', 'gold', and 'tiger', refer to the same natural kind in every possible world in which that kind is instantiated. Famously, Putnam proposed that 'water' refers in every possible world to samples of "the same liquid as" the sample that was initially baptized with the term 'water'. To qualify, in the case of water, a sample has to have the same microstructure, which is captured by the chemical formula H_2O. The argument for this conclusion was based on his famous Twin Earth thought experiment, which posits a planet just

---

[15] This is the view of cladistic taxonomy, but similar problems arise on other taxonomic accounts; see LaPorte (2004, 52–60) for further justification.

like ours in all respects but one. In this counterfactual situation, there is a substance with all the same macroproperties as water but a different microstructure, call it XYZ, which is what covers 71% of Twin Earth's surface, comprises most of the volume of Twin Earthians' bodies, is used to make Twin Earthian lemonade, and so on. According to Putnam, our term 'water' fails to refer to the substance that fills the oceans on Twin Earth, and we should conclude that there is no water on Twin Earth. All and only samples of the natural kind *water* have the property of possessing the microstructure H_2O. This is not only true in the actual world but in every possible world – hence the conclusion that water is necessarily H_2O. Or, to put it in terms that I have been using so far, every sample of the natural kind *water* necessarily possesses the property of having the microstructure H_2O. Such necessities are supposed to hold of natural kinds but not of nonnatural kinds. To be sure, we might discover that the liquid that we have been drinking, swimming in, watering our plants with, and so on, was characterized by a different chemical formula and our current knowledge of chemistry may turn out to be mistaken in various ways. But given that our best theories of chemistry are correct and water is actually H_2O, then water is necessarily H_2O.

Initially, Putnam's work was widely interpreted as having demonstrated, based on the semantics of natural kind terms, that natural kinds have essences, specifically in the sense that they are necessarily associated with certain properties. But upon closer inspection, Nathan Salmon (1981) argued persuasively that the metaphysical thesis about natural kinds does not strictly follow from the semantics of natural kind terms. Salmon showed that one cannot derive the metaphysical conclusion from a consideration of the semantics of natural kind terms without smuggling in certain metaphysical assumptions.[16] With the benefit of hindsight, this result may not seem too surprising, since it is doubtful that a substantive metaphysical thesis could be derived from a semantic one. How could an account of how we apply our general terms somehow yield a conclusion concerning the furniture of the universe? In the above example, the metaphysical assumption is most clearly evident in how we interpret the relation "the same liquid as." If what it is for a sample to be a sample of the same liquid as the initial sample of water is for it to have the same microstructure, then one can show that water is necessarily H_2O. But some philosophers have argued that what it is to be the "same liquid as" the familiar substance

---

[16] For a clear account of the relation between the semantic thesis and the metaphysical one, see Mumford (2005).

water might not be a matter of sharing the same microstructure. Indeed, our judgment as to what it is to be a sample of water might vary from context to context (cf. Mellor 1977; Zemach 1976). If, instead, we consider that what it is to be the same liquid as a certain sample of water is to possess certain macroproperties rather than share a certain microstructure, then Putnam's conclusion that there is no water on Twin Earth and that water is necessarily H₂O does not follow from his thought experiment. Still, it may be said that this assumption is itself warranted since it rests on a strong intuition that the same-liquid relation as applied to natural kinds like water does indeed pick out samples with the same microstructure, whereas this is not the case for nonnatural kinds. I will now consider the plausibility of this way of defending Putnam's conclusion.

The second modal thesis about natural kinds, which states that they are linked as a matter of metaphysical necessity to a certain collection of properties, cannot be supported by considerations based solely on the semantics of natural kind terms. Can it be supported by something like a bare intuition about such kinds themselves? On this way of seeing things, what sets apart paradigmatic natural kinds, such as *proton, gold, water*, and *tiger*, from putative nonnatural kinds, such as *air, mud, weed*, and *bug*, is (at least in part) their modal status. Intuitively, the former are the kinds they are necessarily and would be so in every possible world, whereas the latter are the way they are contingently and would not be associated with the very same set of properties in other possible worlds. Does this intuition stand up to scrutiny? Before answering this question, it is necessary to say more about the semantics of natural kind terms. Recent work in the philosophy of language has established that the semantics of general terms is importantly different from that of singular terms.[17] Specifically, it has now been widely accepted that it is problematic to consider natural kind terms to be "rigid designators," terms that refer to the same thing in every possible world (in which that thing exists). Unlike proper names and other singular terms, there is no straightforward way of applying the notion of rigid designation to general terms, including natural kind terms. The chief difficulty with the claim that natural kind terms refer to the same thing in every possible world can be presented in the form of a dilemma. On the one hand, if one means by the "same thing," the individual members of the kind, then the claim is surely false. Take a natural kind term like 'tiger',

---

[17] Though this observation was made by Donnellan (1983), its significance was not sufficiently appreciated until much later. See also LaPorte (2000), Schwartz (1980, 2002), Soames (2002), and Sterelny (1983).

whose extension in the actual world includes the set of all tigers (whether living or not). Now consider a possible world in which all the actual tigers were replaced by their *possible siblings*, individuals who were not conceived in this world but could have been. It seems uncontroversial to say that our term 'tiger' when applied to this possible world would pick out these individuals though none of them are identical to the actually existing individuals in this world. Therefore, the term 'tiger' does not pick out the same individuals in every possible world. On the other hand, if one means by the "same thing" the same *kind* of thing, then this appears to be a trivial demand that may be satisfied by any general term whatsoever, not just natural kind terms. Consider some putative examples: terms like 'air' or 'mud', which essentialists hold are not natural kind terms. It is reasonable to say that the term 'air' picks out the same kind in every possible world – namely, a kind with the *very same properties* that air has in this world. Those who share Twin Earth intuitions about water should surely also assent to the claim that if Twin Earth had an atmosphere composed predominantly of helium, we would say that there was no air on Twin Earth (not that air on Twin Earth consists of helium). It is intuitively plausible that the term refers in every possible world to a substance with roughly the same composition that air has in this world (namely, 78% nitrogen, 21% oxygen, 1% argon, etc.). Hence, even a term for a nonnatural kind can be said to be associated necessarily with the same kind. Now, essentialists might counter that it is the necessary association with a set of *microproperties* that distinguishes natural kind terms from nonnatural kind terms, and that there is no single chemical compound associated with the kind *air* (in this world or any possible world). But if so, then the difference is not about modality but about an association with a single set of microproperties, an association which is supposed to hold in this world as well as in every possible world. To make this point clearer, consider a category that is even more obviously an example of a nonnatural kind, say, *object with a mass of 67 kg*, and let us coin a term to designate this artificial category, say, 'owamosk'. Then 'owamosk' could be said to pick out the same kind of thing in every possible world. What kind of thing is that? The kind associated with the property of having a mass of 67 kg. Thus, terms for nonnatural kinds, no less than terms for natural kinds, can be said to pick out the same kind of thing in every possible world.[18]

---

[18] LaPorte (2000, 305) allows that: "No good reason has been identified for withholding rigid status from terms for artificial kinds." But he maintains that some general terms are not rigid – for example, 'the insect species that is typically farmed for honey'. I think this point is debatable but it is

Although this fact about natural kind *terms* has now been noted by a number of philosophers and has gained wide acceptance, there is a corresponding fact about natural kinds themselves that has not been appreciated. Just as there is nothing to prevent general terms that do not refer to natural kinds from picking out the same kind across possible worlds, there is nothing to prevent nonnatural kinds themselves from being associated with the same property or set of properties across possible worlds. The difference between natural kinds and nonnatural kinds is not that there is a metaphysically necessary connection between natural kinds and their associated properties, for the same could be said of nonnatural kinds, such as *air* and *mud*, or indeed the made-up kind *owamosk*. Even if we accept, as a basic philosophical datum, the intuition that the kind *water* is necessarily a compound with a certain chemical composition, it does not seem possible to deny the intuition that the kind *air* is necessarily a mixture of gaseous substances in certain proportions. Still, is it not the point that water is a single substance, while air is not a single substance but a mixture of such substances? Yes, but this difference between water and air is not a difference in their modal properties. It is a difference in the properties associated with the kind in the actual world (properties that we continue to associate with that kind in other possible worlds). There is no basis for distinguishing among natural and nonnatural kinds on the grounds that the former but not the latter are intuitively associated with the same set of properties in every possible world. Though a case may be made for denying that a natural kind can consist of a mixture of substances with distinct chemical compositions, that claim is not a claim about the modal properties of natural kinds. Moreover, the modal necessity thesis about general terms and categories even holds for categories that are relational in nature. For instance, the term 'atmosphere' can be taken to refer in every possible world to whatever gaseous mass happens to surround a certain planet, and the kind *atmosphere* may also be associated with the property of being a gaseous mass that surrounds a planet. And that very property, of being a gaseous mass that surrounds a planet, is a property that is intuitively associated with the kind *atmosphere* in every possible world (if Twin Earth were not surrounded by a gaseous mass, then we would presumably say that it has no atmosphere).

In this section, I have examined the claim that real or natural kinds are characterized by having certain modal features. I distinguished two modal

enough for my purposes that not all rigid general terms are natural kind terms, and that some terms for non-natural kinds can also be considered rigid.

theses commonly associated with natural kinds, essentialism about kind membership and essentialism about kinds themselves. After finding the first thesis to be problematic, especially as applied to the most fundamental kinds in nature, I argued that we have no reason to think that it distinguishes natural from nonnatural kinds. The second thesis, which states that natural kinds are those necessarily associated with a certain set of properties, turns out to be trivially satisfied, not just by natural kinds but also by clear examples of nonnatural kinds.

1.6 INTRINSICALITY

There is considerable debate concerning the notion of an intrinsic property, including the correct way to distinguish intrinsic from extrinsic properties. My aim in this section is not to try to resolve this debate. My more modest goal is to argue that, on some of the most plausible accounts of what it is for a property to be intrinsic, there are no good reasons for restricting the properties associated with natural kinds to intrinsic properties.

Intuitively, the distinction seems clear enough: Intrinsic properties are ones possessed by an individual independently of anything else, whereas extrinsic properties are ones an individual has by virtue of its relations to other things in the universe. To take just one example, *having a unit positive charge* is an intrinsic property of a proton, whereas *being bonded to an electron* is an extrinsic property, which is possessed by some though not all protons. Clearly, these two notions, intrinsic and extrinsic, are mutually exclusive. But it turns out to be difficult to formalize this distinction in a rigorous way and several attempts to do so have fallen foul of paradoxes and other difficulties. I will not attempt to summarize the entire debate, but will merely outline a few notable attempts in order to illustrate some of the difficulties. Following Kim (1982), a number of authors have tried to formalize the notion of an intrinsic property by saying that a property P is intrinsic if and only if it is compatible with *loneliness*; namely, the property of being all alone in the world.[19] That is, the only properties that are intrinsic are the ones that a thing could have if it were the only thing in existence in the universe. This accords well with the above examples, since a particle could have a unit positive charge while being the only thing in the universe, but it could not be bonded to an electron if there were nothing else (in this case, at least one electron). But this formulation is

---

[19] This is not exactly Kim's account, but for my purposes it will serve well enough to indicate the difficulties.

susceptible to the following problem, which was pointed out by Lewis (1983). Consider the property of *loneliness* itself: Is it intrinsic or extrinsic? Intuitively, it would seem as though it must be extrinsic since it has to do with whether other things exist and a thing's having it is dependent on whether something else exists. However, given the above definition of intrinsicality, a property is intrinsic if and only if it is compatible with loneliness, and surely loneliness is compatible with itself, so it is intrinsic. Hence, this account seems to give the wrong answer when it comes to the property of *loneliness*.

This defect might be thought to be capable of a quick fix, as follows. A property P is intrinsic if and only if both the presence and the absence of P are compatible with both loneliness and the absence of loneliness. That is, a property is intrinsic to something in case that property could be possessed or not in both a lonely world and in a world populated by other things. This avoids the previous problem, since loneliness is not compatible with the absence of loneliness (and the absence of loneliness is not compatible with loneliness). But it has problems of its own. As Vallentyne (1997) points out, the property of *being the only red object*, which seems clearly extrinsic since it is dependent on whether other (red) objects exist, comes out to be intrinsic on this definition. That is because both its presence and its absence are compatible with both loneliness and nonloneliness. Something could be the only red object in a lonely universe (if the object is red) as well as in a nonlonely one (if it is red and there are no other red objects), and it could be that something is not the only red object both in a lonely universe (if the object is not red) and in a nonlonely universe (if it is not red, or it is red and there are other red objects).

Now these difficulties might seem to be of a technical nature and might appear to admit of suitably technical remedies, but Vallentyne (1997) argues persuasively that the basic problem is that we are trying to use a logical notion of compatibility or independence to capture a different notion of independence, one that is of a metaphysical nature. An intrinsic property is one that is independent of anything else, but not in a strictly logical sense. As Vallentyne puts it, the problem with this analysis (and similar ones based on purely logical notions) is that "It fails to capture the idea that an object can cease to be the only red object in the world by the 'mere addition' of a red object to the world" (1997, 211). Vallentyne tries to fix it by adopting a different approach; namely, by using the notion of a contraction of a given world, "which is to be understood as a world 'obtainable' from the original one solely by 'removing' objects from it" (1997). He then defines an intrinsic property, roughly, as one that an object would have in every contraction of a world, where this includes even removing the natural laws that obtain at that world. But this seems to work only if we understand the notion of what it is to remove everything from the world except a particular object, which may presuppose that we already know what is intrinsic to a given object. (For example, does it include its dispositional properties, properties whose natures depend on natural laws, or the natural laws themselves?)

Another prominent attempt is due to Langton and Lewis (1998), who also try to fix the second definition, but do so by ruling out disjunctive properties. According to them, the reason that the property of *being the only red object* trips up the second definition is that it is implicitly disjunctive.[20] As they freely admit, their analysis relies crucially on the notion of a 'natural property', for a disjunctive property is not merely one that can be expressed as a disjunction, since any property can be so expressed. For example, the property *P* is equivalent to *(P & Q)* V *(P & not-Q)*. Rather, a disjunctive property on their account is one that can be expressed as a disjunction of *natural properties* and is not itself a natural property. Once the second definition is restricted to nondisjunctive properties in this sense, they claim that it delivers the goods. But since the notion of a natural property is something that is in need of clarification, especially if our task is to get clear on the notion of a natural kind, that assumption spoils the account for use in this context. It may not be sufficient for something to be a natural kind for it to be associated only with natural properties (a motley collection of natural properties may not add up to a natural kind), and it may not be necessary either, but it would be problematic to use the notion of a natural property to ground the notion of a natural kind.

If it is so difficult to specify what an intrinsic property is, then it is hard to say whether we should restrict natural kinds to those associated with intrinsic properties, or impose this as a condition on the essential properties associated with natural kinds. But setting this aside, and relying on our common understanding of what an intrinsic property is, do we have any reason to believe that only such properties should qualify as the essences of natural kinds with natural kinds? It might seem so, since what natural kind an individual belongs to should surely be a

---

[20] They analyze it as follows: 'red & not accompanied by another red thing', which is equivalent to: 'red & (non-red V not accompanied by another red thing)', which is equivalent to: 'not [non-red V (red & accompanied by another red thing)]'.

function of how that individual is in and of itself, but is there a more compelling reason for thinking so?

One reason that essentialists may give for associating natural kinds solely with intrinsic essences has to do with their commitment to modal necessity.[21] The idea is that if a kind K is associated with a set of essential properties $P_1, \ldots, P_n$, at least one of which is not intrinsic, then the possession of that extrinsic property, P_i, is dependent on the existence of something else – call it x. But then K cannot be instantiated unless x exists. However, this would not undermine essentialism about kinds (the second modal thesis, discussed in section 1.5.), since K can still be necessarily the kind it is; that is, the kind associated with the same set of properties in every possible world. It just means that K cannot be instantiated in those worlds in which x does not exist. There is no obstacle to including nonintrinsic properties among those essential to a natural kind, as long as one acknowledges that the kind cannot be instantiated in those worlds lacking the conditions necessary for satisfying the extrinsic property. Do extrinsic properties undermine essentialism about kind membership (the first modal thesis)? Here, the matter may seem trickier, since it could be argued that an individual member i_1 of kind K would cease to belong to K in a world in which x does not exist. But an essentialist about kind membership would surely insist that i_1 would not exist unless it belonged to K, so a world in which x does not exist is a world in which i_1 does not exist either. Hence, even if we had no qualms about the two modal necessity claims discussed in the previous section, they pose no obstacle to natural kinds being associated with extrinsic properties. It should therefore come as no surprise that some essentialists are quite happy to include extrinsic properties among the properties associated with natural kinds. Many essentialists about individuals, particularly human individuals, are origin essentialists, holding that what is essential to an individual human being is that individual's origin. Furthermore, some essentialists about kinds, particularly biological species, are origin essentialists, holding that the essence of a given species consists in being descended from a certain ancestral species and occupying a certain place in the phylogenetic tree (Griffiths 1999; LaPorte 2004; Okasha 2002). In this case, the only property associated with the kind is an extrinsic one, since such essentialists do not spell out other, intrinsic properties required for membership in a species. Each species S derives essentially

---

[21] See, e.g., Ellis (2001, 19–21), especially #4.

from certain ancestral species; in worlds in which its ancestral lineage does not exist, species S would simply not exist.[22]

Are there other grounds for associating natural kinds exclusively with intrinsic essential properties (again, relying on a rough-and-ready understanding of intrinsicality, in the absence of a satisfactory analysis)? As stated above, if a property P_I is extrinsic to some individual i_I, then its possession is dependent on the existence of, and perhaps relation[23] to, something else, x. That means that if x goes out of existence or alters its relation to i_I, then i_I will no longer possess P_I. Moreover, if P_I is one of those properties determining whether i_I belongs to some kind K_I, then i_I will thereby no longer belong to K_I. Does the presence of the extrinsic property P_I disqualify K_I from being a natural kind? It might be said that whether an individual belongs to a natural kind or not should surely be a fact about that individual itself, not a fact that is dependent on other things or relations to other things. This can be brought out by harking back to a traditional Aristotelian version of essentialism, according to which the essence of a thing is understood in terms of what it is to be that thing. It might be claimed that the "what-it-is" (the literal translation of Aristotle's phrase that is traditionally rendered by "essence") of something must be intrinsic to it and cannot be dependent on something else. This line of thought might also be encouraged by reflecting on the toy examples of extrinsic properties commonly used by philosophers, such as the property of *being within one mile of a rhododendron*, which strengthen the impression that extrinsic properties are trivial. But consider the fact that contemporary physical theory holds that it is a law of nature that quarks cannot be isolated but must always be bonded with other quarks in a hadron. Surely, this is a fact about quarks themselves, but the property of *being bonded to another quark* is nevertheless an extrinsic property (according to a number of the philosophical analyses discussed above, as well as according to our common-sense conception). It may be that this property of quarks is derivative of a more basic intrinsic property to

[22] To what extent would the rest of the phylogenetic tree have to be identical in another possible world for us to say that S exists in that world? And, are there no restrictions on the synchronic characteristics (i.e., morphological, physiological, and other features) of members of S for S to count as the same species? In my opinion, these questions have not received enough attention from origin essentialists about species and may pose a serious threat to the position. But the position cannot be ruled out simply on the grounds that essential properties cannot be extrinsic.

[23] Is the intrinsic–extrinsic distinction the same as the relational–non-relational one? They seem closely related, though Weatherson (2006) argues that the latter are properties of concepts rather than properties.

be explained by superstring theory, but that is a matter for empirical investigation and the alternative cannot be established by philosophical conjecture. Or consider another example: According to the Pauli exclusion principle, no two identical fermions can share the same quantum state (e.g., no two electrons can be in the same quantum state). The property possessed by each fermion that it cannot occupy the same quantum state with an identical fermion, which follows from a law of nature, is surely not intrinsic since it makes reference to other fermions. Hence, some extrinsic properties may indeed be fundamental properties possessed by individuals that hold by virtue of certain basic laws of nature. Since the kinds *quark* and *fermion* would seem to be no less real for including extrinsic properties among their fundamental properties, and since it is at least possible that these properties do not follow from more fundamental intrinsic properties, we can once again question the connection between a kind's being natural and that kind's members possessing their kind-determining properties intrinsically. The same applies to biological species, which, according to some conceptions, are the kinds they are by virtue of relational properties of descent from a certain lineage. As already mentioned, this extrinsic account of species has been embraced by some essentialists, indicating that some of them at least do not regard the intrinsicality of essential properties as a condition on natural kindhood.

In this section, I have sampled a few notable attempts to give an account of what it is for a property to be intrinsic, all of which seem prone to objections. In particular, one prominent analysis helps itself to the notion of a 'natural property', which is precisely part of what is at issue when one is attempting to give an account of a natural kind. But even if one sets these qualms aside and adopts a rough-and-ready understanding of what an intrinsic property is, there do not seem to be any general grounds for restricting the essential properties associated with natural kinds to intrinsic ones, even by essentialist lights. Furthermore, I have argued that some of the most basic of all kinds (*quarks, fermions*) possess extrinsic essential properties that appear to be a direct result of fundamental laws of nature. Finally, many essentialists about biological species regard the property essential to any given species to be an extrinsic one; though the specifics of the position may be problematic, the position is not incoherent on account of positing an extrinsic essential property. Not only do these essentialists tolerate extrinsic properties among the essential ones, but they also posit exclusively extrinsic essences.

1.7 MICROSTRUCTURE

The next essentialist requirement to consider is that natural kinds must be associated with microstructural properties. Many, though not all, essentialists now hold that the only real properties and kinds are the microstructural ones. Originally, essentialists did not insist that the kinds themselves were necessarily microphysical. Putnam (1973) and others allowed for macroscopic natural kinds such as biological species, but posited that the properties characteristic of such kinds would be microstructural, in this case the structure associated with the DNA sequence characteristic of a particular species. More recently, it has become increasingly clear that there are no microstructural properties that all and only members of a given biological species share, since there is no unique DNA sequence that characterizes the members of any given biological species. Indeed, it is not clear that there are any macroscopic kinds that can be characterized solely in terms of microstructural properties. Hence, many essentialists now tend to limit natural kinds to the microphysical ones themselves: chemical compounds, chemical elements, elementary particles, and so on (Ellis 2001). On this view, call it "microphysical fundamentalism," only the microlevel kinds and their properties really exist. Predicates at the macrolevel may apply to things and group them into categories based on utility, but they do not pick out real properties and fail to carve nature at the joints. Furthermore, such predicates can enter into generalizations, but these are not fundamental laws of nature. The predicates of the special sciences do not pick out properties and kinds, and the generalizations of the special sciences are not statements of natural law. This view is held not only by essentialists but also by some nonessentialist philosophers (e.g., Heil 2003), and it is worth discussing in its own right, not just as a central thesis of many essentialist positions.

Why privilege the *micro* over the *macro*? There is clearly an asymmetry between the smallest realm in nature and all larger realms, since the entities of the former *constitute* the entities of the latter. Kitchen tables (at least some of them) are made up of cellulose molecules, but cellulose molecules are not made up of kitchen tables. If we take the fundamental particles out of the universe, everything else goes with them, including kitchen tables and houseflies, babies and bathwater, asteroids and supernovae. By contrast, the fundamental particles could be rearranged in such a way that all the macrolevel entities were destroyed, leaving the particles intact. All this is just to say that macrolevel entities are *nothing but* the fundamental particles that constitute them. That is what gives these fundamental particles and the natural kinds into which they are categorized a special metaphysical status. Call this the "nothing-but argument" for microphysical fundamentalism.

One problem with this view is the possibility that this is not the ultimate level of nature. After all, scientific inquiry over the past century has uncovered a successive series of ever smaller levels of reality: Molecules gave way to atoms, which gave way to electrons and nuclei, which gave way to protons and neutrons, which gave way to quarks. It would take more energy than is currently available, or indeed may ever be available, to conduct the scattering experiments needed to determine whether quarks have inner structure. It may be that there are further levels of structure at yet smaller scales, which we will not, and perhaps cannot, uncover. Metaphysicians who regard the natural kinds to be identical with the kinds of elementary particles (and the kinds of their interactions) may continue to insist that the smallest scale of reality would determine the ultimate natural kinds, even though we may never manage to identify that scale. They may be ready to admit that the real properties and natural kinds might remain unknown, and perhaps unknowable. But it seems quite possible that there is no ultimate level. Reality may not bottom out in a bedrock of fundamental particles but may instead consist in a bottomless pit of ever smaller levels, what Lewis (1986, 30) calls "infinite complexity" and Block (2003, 138) refers to as the possibility that there is "no bottom level." Even if that possibility were ruled out by our current physics, our current physical theories may not be the last word on the issue. Philosophers who locate natural kinds exclusively at the smallest microlevel are in trouble if there is no smallest level, since that would mean that there are no natural kinds, according to them. However, not only is it a gamble to equate what is real with what is smallest, but also it does not seem correct to say that the existence of a never-ending series of levels would in fact imply that there are no natural kinds. The point is not merely that metaphysicians who equate what is real with the smallest microlevel are taking a risk. Rather, the more telling point is that if there were no ultimate level it would seem wrong to conclude that there were no natural kinds and no real properties. But if that is so, then it follows that it is misguided to equate reality with the smallest level, even if there is a smallest level.[24]

---

[24] For a thorough discussion of the issue, see Schaffer (2003), who appears to be making a similar point at the end of his paper. Ladyman and Ross (2007, 178) take the possibility that there is no fundamental level seriously, but Heil (2003, 173–178) is skeptical.

Another objection against the view that the only real properties and natural kinds are microphysical proceeds by noting an apparent inconsistency in this view. Some philosophers who insist that the natural kinds are microphysical nevertheless seem to countenance the reality of levels above the very smallest level. These philosophers speak of particles such as protons and neutrons, as well as of atoms and molecules, as though they too belonged to natural kinds. But it is not clear on what grounds they can do so. Protons are composed of three quarks, two up quarks and one down quark. Philosophers who insist that electrons and protons cancel out kitchen tables and supernovae must, to be consistent, acknowledge that quarks cancel out protons. According to current physical theory, all matter is composed of elementary particles: protons, neutrons, electrons, photons, gluons, and a handful of other particles. Moreover, some of these particles, the hadrons (which include protons and neutrons) are further composed of quarks. Currently, there are thought to be a relatively small number of kinds of elementary particle: six kinds of quarks and their antiparticles, six kinds of leptons and their antiparticles, and several kinds of bosons, which include photons. If microphysical fundamentalists are right to say that higher-level kinds do not really exist, this surely includes anything higher than the *most fundamental* level – namely (given our current physical theories), quarks, leptons, and bosons. This would leave us with a small number of natural kinds, perhaps a few dozen in total. That may not be a liability in and of itself. However, it is not an attractive prospect when one considers some of the reasons given by essentialists and others for positing natural kinds in the first place. Armstrong, who is of the microstructural persuasion, even though he is not an essentialist, thinks that natural properties and natural kinds enter into *natural laws*. As mentioned in section 1.2, according to Armstrong (1992/1997, 164–165), a natural law, which is the "ontological correlate" of a true law-statement, holds in virtue of "a link between properties." But if the only real properties are those possessed by the elementary particles, then only the laws associated with those properties have ontological correlates. The rest do not correspond to anything ontologically real. And what goes for natural laws goes also for causation, so that purported causal relations that do not obtain by virtue of real properties are not real instances of causation. Hence, real properties and the natural kinds associated with them cannot serve to ground such laws as those of thermodynamics, or optics, or solid-state physics, not to mention the laws of chemistry, biology, and so on. Unless these laws are reducible to those of fundamental microphysics, they cannot be considered to be laws at all.

This objection ought not to seem too surprising, since one widespread argument in favor of microphysical essentialism, or indeed any view that regards only microphysical kinds and properties to be real, is that any admission that macrolevel kinds and properties are real in their own right is liable to lead to causal overdetermination. The claim is that if we posit real properties and natural kinds at all but the microlevel, then causation at higher levels is surely superfluous. The most common version of this argument deals with mental causation: If brain states cause other brain states by virtue of possessing neural properties, then there is no causal work left for mental states to do. Any causation between mental states would be a case of overdetermination. Causation can only take place at the neural level, where instances of various neural properties participate in causal relations with instances of other neural properties. As I have urged above, to be consistent, this argument must be applied to all levels above the minutest level, which means that, strictly speaking, there are no neural properties either, just properties of elementary particles that compose the neurons and other components of the brain. But the problem with this position is that it strips properties from all levels above the smallest one. And this, in turn, means that all purported instantiations of natural laws and all alleged instances of causation at these levels are illusory. The causation is really taking place "further down," and the only laws that operate are those of elementary particle physics. Hence, the vast majority of scientific predicates do not pick out real properties and natural kinds, and the vast majority of scientific law statements do not correspond to natural laws. Without denying the problem of overdetermination for opponents of the microstructural position, I am arguing that the microphysical fundamentalists have a problem of their own – namely, the illusoriness of higher-level properties and the epiphenomenalism of higher-level causation. (In section 3.3, I will have more to say about how one might avoid this dilemma.)

I have argued in this section that the view that the only natural kinds are microphysical ones is open to some damaging objections. One argument for adopting this view, which points to the causal exclusion of the macro, leads to the conclusion that the macro is causally inert and epiphenomenal. That would make all special-science kinds ersatz and all special-science causation illusory, indeed even kinds and causation at the level of protons and atoms. Another problem with the microphysical fundamentalist position is that it would be in trouble if it were the case that there is no smallest level. But not only is it risky to rest the position on the assumption that there is a bottom level, but also a discovery that there is no bottom level would surely not be a discovery that there are no real properties and natural kinds. If so, then it would be a mistake to identify natural kinds with the smallest microphysical kinds.

1.8 CONCLUSION

Two philosophical accounts of natural kinds have been the focus of this chapter. These two accounts are compatible; indeed they are often held together. The first, metaphysical Realism, holds that natural kinds are to be identified with universals, abstract entities over and above the members of the kind or the set of its actual members. The position has problems, particularly in articulating the relationship between universals that correspond to kinds and those that correspond to properties. But the issues raised by this position will not be further discussed in this book. Accounting for the metaphysical status of kinds is a perplexing question, to be sure, but I will not attempt to address it since the focus of the book is on a different question – namely, the distinguishing features of natural kinds or what differentiates natural from nonnatural kinds. As I mentioned in section 1.3, there is a difference between two questions: (i) *what* are natural kinds, metaphysically speaking, and (ii) *which* kinds are natural kinds? The second philosophical account, essentialism, attempts to supply an answer to the latter question. It specifies certain criteria that natural kinds must satisfy, but I have argued that the criteria adduced by essentialists are not adequately motivated and some are incompatible with the findings of contemporary scientific theory about paradigmatic natural kinds.

In the following chapters, I hope to be able to articulate an alternative to the essentialist view of natural kinds which preserves the essentialist emphasis on discoverability by science but rejects the other essentialist conditions on natural kinds. The investigation will be guided by a range of actual examples from science, in an attempt to determine what are the characteristic features of natural kinds and which categories correspond to natural kinds.

CHAPTER 2

The naturalness of kinds

2.1 NATURAL KINDS AND EPISTEMIC KINDS

In the previous chapter, I argued that metaphysical Realism is powerless to distinguish natural from nonnatural kinds, and that the criteria put forward by essentialism are not suitable for doing so. This leaves us with a philosophical problem: What is the difference between natural kinds and nonnatural kinds? Surely, not any old category of things nor any random collection of entities cobbled together constitutes a natural kind; but if not, on what grounds are we to decide which categories do and which do not? If the aim is to discover the categories that really exist in nature, then the obvious place to look is science. After all, that is the enterprise dedicated to discovering the truth about nature. Science is concerned not just with ascertaining particular facts about individual things but more centrally with classifying those individuals into categories and making generalizations about them. The theoretical terms of science purport to pick out properties and kinds. If one is a realist about science, then one will think that scientific categories aim at identifying the kinds that really exist in the world, as opposed to inventing categories that will merely serve certain parochial practical ends. At this point, I will not attempt to argue explicitly for scientific realism or justify it fully, but in Chapter 6, I will try to make the case for realism, or a particular brand of realism about natural kinds, more plausible. For the moment, I will take a realist stance towards science and argue that natural kinds will correspond, at least in the limit of inquiry, to the categories employed by our considered scientific theories.[1] The idea that natural kinds are discoverable by science is not controversial among those philosophers who believe in natural kinds. What may be more

---

[1] I will not attempt to defend scientific realism comprehensively against instrumentalism or constructive empiricism; for a recent defense of this kind, see Chakravartty (2007).

contentious is the claim that *all* the categories of science correspond to natural kinds. Essentialists and other realists about natural kinds are typically quite willing to say that all natural kinds are discoverable by science, but not that all scientific categories align with natural kinds. In this and subsequent chapters, I will attempt to justify both claims. In this chapter, I will also discuss whether natural kinds may be found outside science and consider whether some folk categories might also correspond to natural kinds.

Some philosophers have drawn a distinction between natural kinds on the one hand and "epistemic kinds" or "investigative kinds" (Brigandt 2003; Griffiths 2004) on the other, and have claimed that many scientific kinds are not natural kinds but belong to the latter category, particularly the kinds drawn from the special sciences. Griffiths (2004, 907) distinguishes between natural kinds and investigative kinds on the grounds, in part, that the latter are subjects of an "open-ended investigation." Of course, there is no guarantee that any particular scientific category introduced in the course of inquiry will remain part of our permanent theory of the world. If investigative kinds are thought to differ from natural kinds for that reason, then we should refer to them as "tentative" natural kinds or "presumptive" natural kinds. If, however, they are thought to fall short of being natural kinds because they do not satisfy some further condition, then I will argue that there is no principled reason to impose a further condition on scientific categories, beyond those that such categories satisfy in order to be accepted as part of our established scientific theories. In other words, my claim is that natural kinds *are* investigative or epistemic kinds, in the sense that they are the categories revealed by our systematic attempts to gain knowledge of nature. Since science provides us with the best insight into the kinds that exist in nature, all the categories of science can be corrigibly considered natural kinds – at least, until such time as they are rejected in favor of others (and these latter categories ought then to be considered natural kinds). In light of the fact that the essentialist understanding of natural kinds was found to be defective in the previous chapter and that some of the main philosophical contributors to the debate about natural kinds have not been essentialists (as I shall attempt to show), rather than reject the term "natural kind" altogether, I would advocate retaining that term and not introducing a new term. In Chapter 1, I discussed the essentialist view that natural kinds should be discoverable by science. I will begin this chapter by re-examining that requirement and though I will endorse it, I will also modify and qualify it in certain ways. The rest of this chapter will be devoted to presenting an alternative account of natural kinds. Then, in Chapter 3, I will directly address the issue of whether there are natural kinds in the nonbasic or "special" sciences.

2.2 DISCOVERABILITY BY SCIENCE

One commonly mentioned condition on natural kinds, by essentialists and nonessentialists alike, is that they should be discoverable by science. A closely related desideratum is that the properties associated with natural kinds ought not to be "superficial" but rather "underlying" properties that are responsible for those superficial properties. The word "underlying" is obscure when used in this context; it seems to trade on an ambiguity between microphysical (since the microlevel is sometimes said to be "under" the macrolevel) and nonsuperficial (since what is "under" is thought to be deeper in a more metaphorical sense). If "underlying" means that the properties are microphysical, then I have already cast doubt on that as a requirement on the properties associated with natural kinds in the previous chapter (see section 1.7). But if it simply means that such properties ought not to be superficial ones, then I have no objection to identifying natural kinds with such properties, since it is presumably those properties (whether microphysical or not) that really make a kind of thing the kind of thing it is. It is generally considered the business of science to conduct inquiries to determine which properties are genuine or real and which merely apparent. Therefore, insisting that the properties associated with natural kinds be nonsuperficial, in this sense, seems equivalent to saying that they should be ones uncovered as a result of a systematic scientific inquiry as opposed to those that first meet the eye. Hence, I will interpret the requirement that the properties associated with natural kinds ought to be nonsuperficial as being roughly equivalent to the criterion of discoverability by science, with some qualifications to be uncovered later in this section.

An important aspect of scientific inquiry when it comes to natural kinds is that the kinds identified by science are considered to be corrigible or revisable. If kinds are found not to correspond to real divisions in nature, they are continually revised and refined until they do so correspond. What is characteristic of this method of revision and refinement is the need to make one's categories delimit a class of things that share not only properties but also interesting or important properties, where these are ones that are projectible, enter into new generalizations, are explanatorily fertile, and generate novel predictions, among other features. Consider a category like *vertebrate*, as applied to biological organisms. This category was initially introduced to apply to all and only creatures with a spinal column, since this feature was thought to be a significant one that distinguished them in an important way from other organisms. With time, as biologists discovered more about such creatures and as the theory of evolution came to be widely established, they came to believe that what distinguished such organisms was primarily a history of descent from a common ancestor. Eventually, this biological taxon, now considered a subphylum, was thought to contain some organisms that *do not* have a spinal column or vertebrae. Nevertheless, scientists continue to regard the subphylum *vertebrate* as a significant group and the organisms that belong to it as having important properties in common. Even though the initial classification was based on a feature that is now thought not to be possessed by all members of the taxon and even though this property is encoded in the very term used to pick out the category, the properties associated with the kind were modified in such a way as to accommodate the prevailing scientific theories and established discoveries about the world. There are features of this example that are atypical, in that the primary basis for classification was once thought to be the phenotypical features or synchronic traits possessed by organisms, whereas it was later held that common ancestry was a more principled way of distinguishing among groups of organisms. But even when scientists do not alter the basic criteria of what they take to be salient, they modify their categories in such a way as to preserve what they regard, from the perspective of their current best theories, as the important differences as opposed to the superficial ones. Hence, a key facet of scientific categories is that they are revisable in light of evidence in such a way as to be associated with important or nonsuperficial properties from the point of view of the science in question. The willingness to revise our categories in this way is a key indicator that the kinds are being altered to conform to nature – in other words, that they are discovered rather than merely invented. We can decide to use some of our words whichever way we wish, but we cannot do the same for our scientific concepts. We may link some concepts indefeasibly to certain definitions (e.g., 'vertebrate' to the property of possessing a spinal column), but if we take the empirical evidence seriously, we will find that we need to introduce different concepts to correspond to real kinds and do the work of science.[2]

---

[2] The phylum Chordata, which includes the vertebrates, has emerged as a more significant kind than the subphylum Vertebrata, which shows that science not only revises the kinds that it introduces but also that it is always introducing new kinds, either alongside or instead of existing kinds.

Perhaps we cannot afford to be so confident that scientific categories always track the divisions in nature and pick out the kinds that are genuinely to be found in the world. When scientists settle on superficial distinctions, these distinctions may be weeded out in the course of further investigation, but what if scientists introduce categories that are stipulative or arbitrary? Worse yet, what if their categories are tainted by certain biases or presuppositions that have little to do with the evidence and more to do with extraneous motivations? It is true that the revisability of scientific categories does not, in and of itself, establish that these categories are discovered rather than invented, since categories may be revised to accord with certain prejudices and preconceptions rather than with nature itself. Moreover, there may be instances in which scientists revise or fail to revise their categories in such a way as not to track natural distinctions. Specific claims of this type have been made, particularly with reference to categories in the social or human sciences, and some of these will be examined in the following two chapters (see especially section 4.7). However, in the absence of specific claims to the contrary regarding actual scientific categories, we can take science to aim (among other things) at discovering the kinds of things that actually exist, and we can conclude that, unless there is a specific reason to regard a scientific category as arbitrary or biased, scientific categories are discovered rather than merely invented and will correspond to natural kinds.

There is certainly no guarantee that our current array of scientific categories will remain in place until the end of inquiry. To say that scientific categories correspond to natural kinds is not to say that they do so at the current stage of scientific investigation, but that they will do so when science has settled on its categories once and for all. Even so, whatever scientific revolutions may yet lie in store for us as we discover more about the natural world, it is unlikely that *all* our current scientific categories will be displaced in favor of new ones. Past scientific theory changes have tended to leave the bulk of scientific categories intact, while ushering in new ones here and there. Even though there may well be upheavals that lead us to discard a significant proportion of our current scientific categories, it is improbable that they will all be abandoned in the course of inquiry.[3]

The idea that science aims to discover the natural kinds, or that scientific categories correspond to natural kinds, is not original with contemporary philosophers. It is a long-standing philosophical claim that

---

[3] I will not try to justify this claim here, but see Khalidi (1998b) for an argument against widespread incommensurability among successive conceptual schemes.

can be traced back at least to Locke, and is prominent in the work of Mill. Since I believe that both philosophers have something to teach us about natural kinds and their place in scientific inquiry, I will selectively highlight aspects of both of their views in the remainder of this section in order to emerge with some further insights about the relationship between natural kinds and scientific categories.

Locke held that general terms "stand for sorts" and that each *sort* (a term he prefers to the Latin terms *genus* and *species*) has an essence that distinguishes it from other sorts and determines its boundaries (1689/1824, III vi §§1–2). Such a *nominal essence*, which can be captured in a definition, Locke distinguished from a *real essence*, elaborating on the difference as follows:

This, though it be all the essence of natural substances that we know, or by which we distinguish them into sorts; yet I call it by a peculiar name, the nominal essence, to distinguish it from the real constitution of substances, upon which depends this nominal essence, and all the properties of that sort; which therefore, as has been said, may be called the real essence: v. g. the nominal essence of gold is that complex idea the word gold stands for, let it be, for instance, a body yellow, of a certain weight, malleable, fusible, and fixed. But the real essence is the constitution of the insensible parts of that body, on which those qualities, and all the other properties of gold depend. How far these two are different, though they are both called essence, is obvious at first sight to discover. (1689/1824, III vi §2)

There is at least one aspect of Locke's view that is in accord with the modern essentialism that was criticized in the previous chapter – namely, his equation of the real essence of a kind with the microphysical constitution possessed by all and only members of a kind. But, in contrast with contemporary essentialists, Locke was profoundly skeptical of the possibility of discovering the real essences of kinds, since he thought that it was beyond human capacity to investigate the microscopic particles (or "corpuscles") of which everything was made. Yet, in certain passages, Locke leaves open the possibility that the properties of the smallest particles might at some future date become known:

But whilst we are destitute of senses acute enough to discover the minute particles of bodies, and to give us ideas of their mechanical affections, we must be content to be ignorant of their properties and ways of operation; nor can we be assured about them any farther, than some few trials we make are able to reach. *But whether they will succeed again another time, we cannot be certain.* This hinders our certain knowledge of universal truths concerning natural bodies; and our reason carries us herein very little beyond particular matter of fact. (1689/1824, IV iii §25; emphasis added)

Locke seems to think that if we were to discover the nature of the particles that make up matter, then we would achieve "certain knowledge of universal truths concerning natural bodies." It is our inability to access these miniscule constituents that prevents us from attaining this knowledge in science (as opposed to mathematics and ethics).

As I mentioned, Locke is one of the main sources for the idea, which I argued against in the previous chapter, that the real or essential properties in nature are located exclusively at the microphysical level. I will not recapitulate my argument here, which casts doubt on the assumption that the properties associated with natural kinds are always microphysical in nature. More crucially, for my purposes here, Locke thinks that "real essences" are to be found not by stipulation but after a process of investigation. Though he was skeptical about the possibility of discovering what really made one kind different from another, and though he seemed to locate these differences solely at the microlevel, he was right to think that kinds were discoverable by science rather than invented or stipulated. Moreover, in some of his more optimistic moments, he seems to have left open the possibility that scientific investigation would one day succeed in discovering the bases of natural kinds.

The claims that natural kinds are discoverable by science and that our catalogue of kinds and the properties associated with them are generally revised in the course of scientific investigation are also emphasized by Mill. He begins by distinguishing "natural" from "artificial" classifications on the grounds that natural classifications are ones that divide objects into groups whose members have a large number of important properties in common:

> The ends of scientific classification are best answered, when the objects are formed into groups respecting which a greater number of general propositions can be made, and those propositions more important, than could be made respecting any other groups into which the same things could be distributed. (1843/1974, IV vii §2)

Regarding a scientific classification system, Mill (1843/1974, IV vii §2) states that "the test of its scientific character is the number and importance of the properties which can be asserted in common of all objects included in a group." By contrast, "artificial" classifications do not group objects into categories all of whose members share many important properties. As an example of an artificial classification, Mill mentions Linnaeus' classification of plants, which was based on the numbers of stamens and pistils in their flowers. The problem with such a taxonomic system, according to him, is that plants with a given number of stamens and pistils do not generally have enough other properties in common to make the classification useful. As Mill puts it: "to think of [plants] in that manner [i.e., as having a certain number of stamens and pistils] is of little use, since we seldom have anything to affirm in common of the plants which have a given number of stamens and pistils" (1843/1974, IV vii §2). Hence, a necessary condition on a natural classification system is that members of its classes must share a *large number of important properties*. Though Linnaeus' system was proposed as a scientific hypothesis, it was rejected in light of further investigation, and the category that it was based upon was revised in such a way as to conform more closely to natural divisions. This constitutes another instance in which scientific categories are revised and refined to pick out natural kinds.

Natural or scientific systems of classification, for Mill, are the ones that identify the "real kinds" in nature. But even though he thinks that all real kinds or natural groups belong to a natural system of classification, not all groups in a natural system of classification are real kinds. To be real kinds, such groups must also satisfy further conditions, besides possessing a large number of important properties in common. Chief among these is that the properties associated with a kind be inexhaustible.[4] As Mill puts it: "Our knowledge of the properties of a Kind is never complete. We are always discovering, and expecting to discover, new ones" (1843/1974, IV vii §4). He also says that "the common properties of a true Kind, and consequently the general assertions which can be made respecting it, or which are certain to be made hereafter as our knowledge extends, are indefinite and inexhaustible" (1843/1974, IV vii §4). In Mill's conception, not only is it the case that natural kinds have a large number of important properties in common, but also the number of such properties is boundless. For example, Mill considers whether there is such a kind as *flat-nosed animal* and says that the answer depends on whether all flat-nosed creatures have an indefinite number of properties in common, beyond the properties that all animals have in common:

To determine whether it is a real Kind, we must ask ourselves this question: Have all flat-nosed animals, in addition to whatever is implied in their flat noses, any common properties, other than those which are common to all animals whatever? If they had; if a flat nose were a mark or index to an indefinite number of other peculiarities, not deducible from the former by an ascertainable law, then out of

---

[4] The second condition on real kinds, that there should be an "impassable barrier" between them, will be discussed in section 2.4.

the class man we might cut another class, flat-nosed man, which according to our definition, would be a Kind. (1843/1974, I vii §4)

Hence, the true mark of a real or natural kind is the inexhaustibility of the properties associated with such a kind. According to this criterion, Mill says, *sulfur* and *phosphorus* are real kinds while *red* and *white* are not.[5]

At first, Mill's criterion seems plausible – namely, that natural kinds correspond not just to those categories associated with a large number of important properties that are discoverable by science, but also to those whose properties are inexhaustible as a result of scientific investigation. Nonnatural categories, by contrast, are ones that have nothing to them besides those properties used to identify them in the first place. Members of the class of all red things are just red and there are no other properties that they all share. Similarly, to revert to a hypothetical example from section 1.5, *owamosk* is not a natural kind, since all that the members of the category have in common is that they are objects with a mass of 67 kg. Mill concludes:

It appears, therefore, that the properties, on which we ground our classes, sometimes exhaust all that the class has in common, or contain it all by some mode of implication; but in other instances we make a selection of a few properties from among not only a greater number, but a number inexhaustible by us, and to which as we know no bounds, they may, so far as we are concerned, be regarded as infinite. (1843/1974, I vii §4)

The latter, but not the former, are considered by Mill to be natural kinds. But despite the initial plausibility of this criterion, it would seem to be seriously flawed, as I shall now argue.

One problem with Mill's criterion of indefinite discoverability was pointed out by early critics of his position. It does not seem possible to say with any definitiveness that the properties that can be discovered of some kind that we identify will be inexhaustible or not.[6] Suppose we find that a number of individuals share a certain number of properties, $P_1, \ldots,$ P_n, and conjecture that they all belong to some natural kind K. Now suppose that we go on to discover that they all share a further number of

---

[5] At times, however, Mill seems not to insist on the inexhaustibility condition on natural kinds. He writes: "Genera and Families [in the Linnaean taxonomy] may be eminently natural, though marked out from one another by properties limited in number; provided those properties are important, and the objects contained in each genus or family resemble each other more than they resemble anything which is excluded from the genus or family" (1843/1974, IV vii §4). For more on "importance," see below.

[6] M. H. Towry (1887, 436) objected by asking rhetorically: "How can we be justified in framing a class upon such a changeable and subjective point as our own ignorance?"

properties, thereby bolstering our conjecture that K is a natural kind. At what point would it be right to conclude that they share an indefinite number of properties and that K is indeed a natural kind? On Mill's behalf, it might be said that even though we may never be in a position to declare with finality that there are an inexhaustible number of properties to be discovered of K, we can at some point safely assume that the number of such properties is indefinitely large. Moreover, even though we may not *know* whether the number of properties is inexhaustible, that does not mean that, whether we know it or not, what distinguishes natural from nonnatural kinds is that the properties associated with the former are indefinite in number. But this response raises another question – namely, why do the properties associated with a natural kind have to be inexhaustibly many? Granted, the main point of positing kinds over and above properties is that they are associated with a multitude of co-occurring properties, but there does not seem to be any need for these properties to be indefinitely large or inexhaustible. I will now propose two reasons why it may seem right to propose this condition on natural kinds. Though neither of these reasons succeeds in justifying the condition, the discussion will help to elucidate the issue.[7]

The first reason for thinking that natural kinds should be associated with an indefinite number of properties has to do with the revisability of concepts corresponding to natural kinds, which was mentioned earlier. Since one characteristic sign that our kind-concepts are being altered to conform to the nature of reality is their revisability in the face of new evidence, it may be thought that this implies that there will be an indefinite number of properties to be associated with each kind. But when we revise our concepts in this manner, some of the properties we once associated with a kind are generally replaced with other properties, in such a way that we are usually not accumulating properties indefinitely. We once thought that all members of the kind *vertebrate* had spinal columns but we now think they share other properties (e.g., descent from a certain common ancestor). For all we know, we might eventually settle (or have already settled) on a fixed and finite number of properties that all and only members of the kind *vertebrate* share. More likely, it is a polythetic kind and there is a loose but finite cluster of properties that

---

[7] There is another problem with Mill's condition, which I will not pursue further. If we accept that the number of properties associated with natural kinds is indefinitely large, this would seem to preclude the possibility of inquiry coming to an end. While I do not have an argument to suggest that scientific inquiry will definitely terminate, I do not think that we can safely assume that it will not.

members of the kind tend to share. Moreover, even though revisability is a sign that our kind-concepts are being made to conform piecemeal to reality, it is possible that we might discover all the properties associated with a kind at the beginning of our inquiry into the nature of that kind. Suppose that after discovering a new kind, we happen to hit upon all its associated properties at the outset of our investigation. Though unlikely, given the way that science typically progresses, such a scenario is not impossible. If this scenario does come to pass, it may make us more skeptical that the proposed kind is a genuine natural kind, but it surely should not rule it out. Hence, potential revisability does not give us a reason for thinking that natural kinds ought always to be associated with an indefinite number of properties.

A second reason that might be put forward for requiring that natural kinds be associated with an indefinite number of properties is that it will prevent bogus or stipulative kinds from creeping in. Perhaps the thought is that if we merely insist that members of a natural kind share a large (but limited) number of important properties, then we could simply introduce a kind that is associated with a large number of properties from the start, and then decree that it is a genuine natural kind. Mill's insistence that members of natural kinds share an indefinite number of properties in common may be thought to guard against kinds that are invented rather than discovered. But insisting that the number of properties be inexhaustible does not seem to be the way of avoiding this. The kind *owamosk* is associated with the properties of having a mass less than 68 kg, having a mass less than 69 kg, having a mass less than 70 kg, and so on. Rather than insist that the number of properties be indefinitely large, what would prevent us from allowing a nonnatural kind to meet this criterion is to insist on two additional conditions that Mill places on natural kinds. The first is the requirement that the discovered properties not follow as a matter of logic from those properties that we already associate with the kind. As Mill puts it in the passage quoted above, the properties in question must "not [be] deducible from the former [property] by an ascertainable law." According to Hacking (1991a, 119), Peirce objected to this requirement on the grounds that part of the point of scientific inquiry is the derivation of certain properties from others as a matter of law. However, Mill can be charitably interpreted as saying that these other properties should not follow by logic alone or in a direct or trivial manner from other properties. Mill (1843/1974, IV vii §2; emphasis added) also says: "The properties, therefore, according to which objects are classified, should, if possible, be those which are *causes* of many other properties."

The second condition, which has already been mentioned, is that the properties associated with a natural kind be *important* in some sense to be further specified. I would argue that if the properties associated with the kind are indeed scientifically important, then even if they are associated with the kind from the outset, it should not disqualify it from being a natural kind. Mill devotes some attention to the issue of "importance," and though he does not delineate the idea clearly, his discussion will serve as a good starting point. He makes two points about importance in this context. The first is that importance is relative to the ends that one intends to accomplish with the classification in question, and is not determined irrespective of a particular disciplinary framework or systematic investigation of some scientific domain. Mill writes:

We said just now that the classification of objects should follow those of their properties which indicate not only the most numerous, but also the most important peculiarities. What is here meant by importance? It has reference to the particular end in view; and the same objects, therefore, may admit with propriety of several different classifications. Each science or art forms its classification of things according to the properties which fall within its special cognizance, or of which it must take account in order to accomplish its peculiar practical end. A farmer does not divide plants, like a botanist, into dicotyledonous and monocotyledonous, but into useful plants and weeds. A geologist divides fossils, not like a zoologist, into families corresponding to those of living species, but into fossils of the palaeozoic, mesozoic, and tertiary periods, above the coal and below the coal, etc. (1843/1974, IV vii §2)

The acknowledgement that "importance" may be relativized to particular sciences and that what may be important from the point of view of one investigation may not be so from another is a salutary proposal. However, Mill does not adequately spell out what would make something important from the point of view of one inquiry or another. To make matters worse, he proceeds to say that there may be such a thing as classification without reference to particular ends or purposes. Thus, his second point is that importance can also be judged outside of any particular context:

These different classifications are all good, for the purposes of their own particular departments of knowledge or practice. But when we are studying objects not for any special practical end, but for the sake of extending our knowledge of the whole of their properties and relations, we must consider as the most important attributes, those which contribute most, either by themselves or by their effects, to render the things like one another, and unlike other things; which give to the class composed of them the most marked individuality; which fill, as it were, the largest space in their existence, and would most impress the attention of a spectator who knew all their properties but was not specially interested in any. Classes formed on this principle may be called, in a more emphatic manner than any others, natural groups. (Mill 1843/1974, IV vii §2)

However, Mill's attempt to characterize importance in a context-free manner does not succeed in doing so in a meaningful way. Giving a class its "most marked individuality" is a vague notion, as is the idea that important properties are ones that would most "impress the attention" of a disinterested spectator. It would seem that Mill has taken a wrong turn here and that he would have been better off grounding the notion of "importance" in the purposes to which science would put the kind in question. We know which categories science considers important and there is widespread consensus on the features that such categories ought to have. They ought to be projectible, enter into empirical generalizations, summarize a wealth of data, feature in explanations, give rise to valid predictions, and so on. As Mill says in one of the passages quoted in the previous paragraph, important properties are "causes of many other properties." Rather than rely on such a precarious idea as the ability to impress a disinterested spectator, Mill would have done better to characterize important properties in terms of their fulfilling a familiar range of scientific desiderata. Moreover, this idea should be qualified by saying that different categories may be found to be important from the perspective of different scientific disciplines or subdisciplines (an idea that will be further justified in section 2.5).

By taking the conceptions of Locke and Mill as our point of departure, we have made some headway on the distinctive features of natural kinds. Each natural kind is associated with a number of properties – namely, the types of properties that are discoverable by science. Not only does science aim at discovering the kinds that exist in nature, but it also revises its categories in such a way as to eliminate categories that do not correspond to genuine kinds of phenomena. Natural kinds ought not to be associated merely with single properties, for the very point of identifying natural kinds is to locate properties that are projectible, pointing to other properties. Nor, therefore, should they be linked to a set of properties all of which are logically deducible from one property. Hence, the properties associated with natural kinds should not follow from a single property in a trivial manner or as a matter of logic, though they will be linked with one another causally or according to natural law. I have found no good reason to require that the properties associated with natural kinds be indefinite in number or inexhaustible, as Mill has it. Should at least some of the properties associated with a natural kind not be, as it were, built into the kind at the outset, but only found to obtain as a result of subsequent scientific investigation? While satisfaction of this criterion may help to reassure us that the kind has not been "cooked up" or gerrymandered, it may not be a good idea to insist on it. After all, there is nothing to prevent us from discovering all the properties that are associated with a kind at the outset of an investigation into that kind. Though science usually progresses gradually in fits and starts – and with plenty of false starts – it is conceivable that we might hit upon all the important properties associated with a kind at the time that we first introduce it. Provided these properties are important ones, this should not preclude the kind in question from being considered a natural kind. As for importance, I have proposed that it be understood in terms of the features that we usually look for in our scientific categories: projectibility, explanatory efficacy, predictive value, and so on. With Mill, we can allow that the ways in which these criteria of scientific importance manifest themselves may vary from discipline to discipline within the sciences, depending on the interests and aims of particular subfields or areas of inquiry. (This would seem to be incompatible with Locke's view, since he holds that real essences of natural kinds are located exclusively at the microphysical level.) It is not clear whether Mill thought that this holds for genuine natural kinds, or only for the broader category of scientific classes or categories contained within a scientific classification system. But I would endorse the former view and will try to justify it further in section 2.5.

2.3 SCIENTIFIC KINDS AND FOLK CATEGORIES

In the previous section, I asked whether all scientific categories correspond to genuine natural kinds, and concluded tentatively that they do, since science aims precisely at uncovering the kinds of things that really exist in nature, at least if one takes a realist stance towards science. But though this may be a sufficient condition on natural kinds, is it a necessary one? Are all natural kinds scientific categories? Could some natural kinds derive not from science but from some other domain of human endeavor? In other words, it might be asked whether scientific categories are the only ones that track the nature of reality and whether it might not be the case that some categories that are not genuinely scientific nevertheless correspond to natural kinds. Insofar as science, in its diversity and with its multitude of branches, is the systematic enterprise dedicated to understanding the natural world, it represents our best bet at uncovering the kinds that are genuinely to be found in nature. The requirement that natural kinds correspond to those categories discovered by science is not insular or scientistic but merely identifies natural kinds with the categories that are posited as a result of a systematic inquiry, as opposed to categories that we might be inclined to conceive as a result of a casual or passing acquaintance with some aspect of reality. It is not out of the question for nonscientific categories, commonly referred to as "folk" categories by philosophers and psychologists, to pick out real kinds of phenomena but I would argue that such categories typically get incorporated in time as part of a scientific discipline or a systematic inquiry.

To assist in the endeavor of trying to ascertain the connection between scientific categories and folk categories, and to determine the relationship that the latter bear to natural kinds, two philosophers will serve as good guides. W. V. Quine, in his classic essay on natural kinds, puts forward a distinctive attitude towards this relationship, according to which folk categories can coexist temporarily with scientific categories, though science eventually renders folk classifications obsolete as it advances. Meanwhile, John Dupré thinks that ordinary language classifications are as legitimate as scientific ones and can exist alongside them indefinitely.

Quine's attitude to natural kinds is somewhat at variance with the approach I have adopted so far. Far from being the categories that science aims to uncover at the end of inquiry, natural kinds are the rough-and-ready categories that we begin with prior to, or at an early stage of, scientific inquiry. For him, natural kinds are grounded in folk classifications; they are sets of things that are all *similar* in some respect. He holds that the notions of *kind* and *similarity* are both somewhat obscure and do not admit of precise definition, though they are inter-defined in various ways. Things that we find similar are placed in sets that we consider to be natural kinds.[8] These similarities are initially grounded in our "innate similarity standards," the quality spacing that is common to members of the human species. But as we discover more about the world, many of these similarities are found to be spurious or not far-reaching enough, so we replace them with similarities that are more in line with the true nature of the universe. "Color is king in our innate quality space,"

---

[8] One notable respect in which Quine's account is at odds with most other philosophical accounts is that he regards natural kinds as extensional rather than intensional entities. He writes: "Kinds can be seen as sets, determined by their members. It is just that not all sets are kinds" (1969, 118). I will ignore this complication in what follows, since I take it as relatively uncontroversial that two (actually) coextensive sets may correspond to two genuinely different kinds.

Quine writes (1969, 127), "but undistinguished in cosmic circles. Cosmically, colors would not qualify as kinds." The sciences replace natural kinds based on color similarities with kinds based on other similarities, and these similarity relations are defined differently in different branches of science. As each science matures, Quine thinks that it will define a precise similarity relation that is applicable primarily to its particular subject matter. Thus, chemistry will define similarity of sample objects by matching their constituent molecules:

Molecules will be said to *match* if they contain atoms of the same elements in the same topological combinations. Then, in principle, we might get at the comparative similarity of objects a and b by considering how many pairs of matching molecules there are, one molecule from a and one from b each time, and how many unmatching pairs. (Quine 1969, 135; original emphasis)

Meanwhile, biology will define similarity of organisms or species in terms of proximity and frequency of common ancestors, or better yet, in terms of common genes (Quine 1969, 137). From these remarks, it seems as if Quine effectively thinks that the notion of similarity will be reduced in each of these cases to some complex relation based on identity (e.g., we might define organism a as being more genetically similar to organism b than to organism c, if and only if a and b have more identical alleles than a and c[9]). This would make similarity drop out as a generic concept in science, being replaced by specific notions of similarity defined in terms of identity of molecules, genes, or similar constituent entities.

Whether or not Quine's hunch about similarity is vindicated or not, there does not seem to be anything to prevent the categories that are devised at the end of such a process from corresponding to natural kinds. Rather than say that the members that belong to such categories are similar, Quine thinks that we would be able to say that they have a certain proportion of identical constituents. But if the categories that are so identified are projectible, have explanatory value, and are otherwise important from the point of view of the relevant science, then they would seem to conform to the notion of natural kind that was discussed in the previous section. Moreover, I suggested in section 1.2 that we should avoid talk of brute similarity among members of a natural kind, and speak instead of members having identical constituents or of sharing properties.

---

[9] This prediction of Quine's has been borne out by various measures of genetic distance that have been developed by geneticists. One of the simplest measures of genetic distance is based on the proportion of shared alleles summed over all genetic loci. This measure can be used for individuals as well as for populations or taxa.

Hence, Quine's conjecture that we will in the fullness of time be able to dispense with natural kinds altogether is not a conclusion that is forced on us by his conjecture that similarity relations will eventually be made precise and relativized to each branch of science, at least not if we understand natural kinds along the lines that I have been proposing.

Quine thinks that natural kinds have their origin in common-sense categories recognized by natural language and will eventually perish with the emergence of an advanced scientific worldview. Though he allows that the notion of a natural kind and the closely related notion of similarity will have a place in the immature sciences, he thinks that these notions will be phased out as the sciences come to fruition. As he puts it in the final sentence of his essay: "In this career of the similarity notion, starting in its innate phase, developing over the years in the light of accumulated experience, passing then from the intuitive phase into theoretical similarity, and finally disappearing altogether, we have a paradigm of the evolution of unreason into science" (1969, 138). However, Quine also states that as long as natural kinds continue to play a role in the immature sciences, these kinds can coexist alongside common-sense kinds. He holds that an "innate similarity notion" can coexist with a "scientifically sophisticated one" and that scientific kinds "do not wholly supersede" the natural kinds that we begin with. As he puts it: "Something like our innate quality space continues to function alongside the more sophisticated regroupings that have been found by scientific experience to facilitate induction" (1969, 129). Quine does not make clear in this essay whether he thinks that our common-sense kinds will eventually be displaced entirely by scientific categories, or whether given certain human concerns, some "intuitive" natural kinds will continue to have a place in our total theory of the world even at the end of inquiry. At times, he implies that the natural kinds embedded in our common-sense concerns will be abandoned altogether, as in the passage quoted about our evolution from "unreason into science." But at other times, he seems to recognize that humans will always have certain mundane concerns and reasons for classifying things that are at variance with scientific classifications. As Quine (1969, 128) phrases it: "If man were to live by basic science alone, natural science would shift its support to the color-blind mutation." He makes it clear in the next sentence that man does indeed live by bread as well as by basic science, thus indicating that our reliance on folk categories may not be wiped out entirely.

I have argued that even if Quine is right to think that similarity will be reduced to the more precise notion of identity, that is no reason to abandon natural kinds altogether. Natural kinds can still be considered to correspond to scientific categories, and the latter can be seen to group together not similar individuals but individuals that satisfy certain precise identity relations or that share certain properties. But what are we to say of nonscientific categories, which, at least at this stage of inquiry, continue to thrive alongside scientific ones: Can they be candidates for natural kinds? Since I have made discoverability by science the central plank of my account of kinds, it may appear that this could not be the case. But, as Quine and many other authors acknowledge, many scientific categories start out as folk categories. Moreover, there are at least some folk categories whose purpose may be to mark distinctions that really exist in nature. This suggests that we might not be able to dismiss such folk categories altogether.

Elsewhere, I have proposed that folk categories can be expected either to coincide with or to be superseded by scientific categories when the purposes for which they are introduced are roughly the same. When they are not, we should not expect them either to coincide or be superseded, but perhaps to coexist alongside scientific categories (Khalidi 1998a). If folk medicine aims primarily to ascertain the real causes of human diseases and the folk are focused on distinguishing kinds of diseases based on their causal properties, then we should expect that folk categories will either coincide with scientific categories, or, when they do not, they will be superseded by them. In some cases, a folk disease that is thought to have certain causes is replaced by one with quite different causes. Our concept of the disease *consumption* was modified greatly when tuberculosis bacteria were discovered, leading us to rename the disease and revise many folk beliefs about its causes (such as the belief that it was caused by vampirism). Meanwhile, a kind-concept like *hysteria*, which was thought to denote a disease primarily afflicting women and involving disturbances in the uterus, proved eventually not to pick out a natural kind of disease and was dropped as a scientific concept. But it may well be that some folk categories are efficacious at treating illnesses and help advance the aim of curing patients and making them feel better, though we have good grounds for thinking that they do not pick out real diseases. That would not be an altogether unprecedented situation, since placebo effects are quite common in medicine. Though in such cases the folk categories may remain in circulation, they should not be considered candidates for natural kinds since their aim is not primarily one of enabling us to predict and explain. If our aim is merely to make patients' lives better, we might continue to employ these categories in a clinical setting and in communication with patients. However, for the purposes of research, we come to recognize that these categories do not conform to real kinds and are merely useful crutches that enable us to accomplish certain fairly narrow goals. Speaking generally, it is quite possible that we might introduce categories that enable us to serve certain desired aims, but do not correspond to the kinds that exist in nature. In such cases, since there is no direct competition between folk categories and scientific categories, we would expect the folk categories to coexist alongside scientific ones. But such folk categories should not be expected to correspond to natural kinds.

At this point, it may be useful to contrast my view with Dupré's on the relation between folk categories and scientific taxonomy. Unlike Quine, Dupré (1999, 461) does not think that folk categories will generally be superseded by scientific ones, and he insists that "folk taxonomies are as legitimate and can be interpreted as realistically, as scientific taxonomies." Indeed, in some of his earlier work, he suggested that the folk classification of whales as fish was not illegitimate, but was rather warranted in certain folk biological contexts (Dupré 1993, 24). However, in more recent work, he has come to revise this judgment, admitting that, in this case at least, the scientific mode of classification has now prevailed over the folk classification scheme. So prevalent has this scientific worldview become that the folk themselves no longer regard whales as fish. Hence, Dupré (1999, 474) concludes: "Regrettably, I have had to admit that whales are not fish, for the sufficient reason that almost everyone in our culture ... agrees not to call them so." But what is missing from this judgment is a consideration of the possible reasons for the purported fact that whales are no longer generally classified as fish. Dupré seems to take it as a brute fact that the folk have deferred to biological practice in this case; indeed, he insists that there "is no good reason" for excluding whales from the category of fish. He thus admits defeat on *de facto* rather than *de jure* grounds: As a matter of fact, the folk have deferred to scientists in this case, but they need not have done so, and if they had not, it would have been quite appropriate to judge that whales are fish in certain folk contexts.[10]

If my proposal is correct, we ought to look for the reasons behind deference and lack of deference in each particular instance. It would be

---

[10] Similarly, LaPorte, who discusses the relationship of folk classification to scientific taxonomy, does not sufficiently investigate the reasons for deference and lack of deference, and ends up sending mixed signals on the issue. He says that the folk regularly defer to the scientists (2004, 31) but also that ordinary usage often persists and parts company with scientific nomenclature (68–69). He also says that vernacular use is often adjusted to conform to science though not always (87–88), but he nevertheless maintains that revision and refinement do seem to be the rule (89–90).

rational for the folk to defer to scientific classification if their purposes coincide with the scientific community, but not if their purposes diverge. If it is indeed the case that the folk have almost universally come to exclude whales from the category *fish* (an assumption I will go on to question below), that is presumably because they share (at least) some of the aims or purposes of scientists in classifying organisms, and these aims or purposes are best served by scientific rather than folk classification. Since this act of classification in biology is based mainly on descent, it would seem as though the folk now also share this interest and have deferred to the scientists at least partly for that very reason. Moreover, in this as in many other biological cases, classification by descent also happens to track important phenotypic features of the organisms involved. Not only are whales not closely related by descent to most of the other organisms we label as 'fish', but also they do not have gills, they give birth to live offspring, and they possess other mammalian properties that fish generally lack. The original classification was presumably based on gross phenotypic features and a broadly shared habitat. Once these properties turn out not to be "important" (in the sense introduced in the previous section), we cease to classify on their basis and seek other properties instead.

So far I have granted Dupré's claim that the folk have largely deferred to the scientists in this case. Whether or not that is so can only be ascertained by a detailed sociolinguistic inquiry that looks at the way in which the relevant terms are used among laypersons and within the scientific community. Though I have not undertaken such an inquiry, there is at least some evidence that this is not entirely the case from lexicography, which tends to summarize ordinary usage. Many standard dictionaries now include two or more entries for the term 'fish', at least one of which refers not to a biological taxonomic category but to the property of being an "aquatic creature" (perhaps prefaced with a parenthetical "loosely" or "colloquially").[11] This provides some reason for thinking that 'fish' is

---

[11] Dictionaries that have a separate entry (or subentry) for the loose usage of 'fish' include the following: *American Heritage Dictionary of the English Language, Merriam-Webster's Online Dictionary* (11th edn.), *Webster's New World College Dictionary* (4th edn.), *Infoplease Dictionary*, and *Dictionary.com*. While the *Oxford English Dictionary* does not have two entries, it clearly distinguishes two senses of the term: "In popular language, any animal living exclusively in the water; primarily denoting vertebrate animals provided with fins and destitute of limbs; but extended to include various cetaceans, crustaceans, molluscs, etc. In modern scientific language (to which popular usage now tends to approximate) restricted to a class of vertebrate animals, provided with gills throughout life, and cold-blooded; the limbs, if present, are modified into fins, and supplemented by unpaired median fins."

equivocal as used in contemporary English, and it is not difficult to see why that would be the case. As Mill (1843/1974, IV vii §2) noted: "Whales are or are not fish, according to the purpose for which we are considering them." His contemporary, Whewell (1847, 459), appears to concur: "A whale is not a fish in natural history, but it is a fish in commerce and law." Hence, it would seem as if there is room for two concepts of *fish*, according to one of which whales are fish and according to the other of which they are not, depending on the purposes for which we want to use these concepts. However, to this, I would add that not all purposes are created equal. Though the folk may have occasion to use the term 'fish' in ways that do not conform to scientific classification, these uses do not appear to be projectible or genuinely explanatory. When the category *fish* includes aquatic animals, such as crayfish, jellyfish, starfish, and some mollusks, as well as whales and dolphins, it ceases to have value as an inductive category. According to the Fisheries Glossary issued by the Food and Agricultural Organization (FAO) of the United Nations, 'fish' used as a collective term includes mollusks, crustaceans, and any aquatic animal which is harvested.[12] But in this inclusive sense, there is nothing more to be discovered about fish. The property of being (capable of being) harvested is a property that was built into it to begin with. It may be objected that this picks out an "important" property (in the sense of the previous section), for the simple reason that anything harvested is subject to laws of supply and demand. But that property is shared with a much broader class of things (commodities), not one pertaining, even loosely, to all and only fish (in the broad sense). Hence, the category *fish*, when interpreted thus is epistemically otiose.

It is instructive to contrast this inclusive use of the term 'fish' with the scientific one. Even though the standard scientific use of the term is itself not free of complication, the category is clearly projectible and has explanatory efficacy. There are over 30,000 species that scientists refer to as 'fish', though they do not belong to a single monophyletic taxon (a taxonomic category that includes all and only descendants of a common ancestor). From the point of view of cladistic taxonomy, which classifies strictly according to descent, there is no taxon that corresponds precisely to the category *fish*. Still, there is enough in common among these species that warrants classifying them in a single category, though there is no property that they all share (which is not also shared by nonmembers). The vast majority of creatures classified as *fish* live in water, breathe with their gills,

---

[12] www.fao.org/fi/glossary/default.asp

are cold-blooded (ectothermic), swim using fins, lay eggs (oviparous), and have scales. These generalizations are not exceptionless; for example, mudskippers live partly on land, lungfish breathe air through their lungs, bluefin tuna are endothermic, and sharks do not have scales. Still, they hold widely enough across the diversity of fish species and the exceptions share enough other properties (including phylogenetic descent) with species that do have these properties to warrant including them in the category *fish*. In terms of shared properties, the category *fish* is a cluster or polythetic kind rather than a monothetic or definable kind (to use terms introduced in the previous chapter), but it is a natural kind nonetheless. Despite the fact that it is not a unitary taxon from the evolutionary or phylogenetic point of view, the category *fish* has undisputed value as an epistemic kind. There are a number of branches of science, such as ichthyology and marine biology, which use this category to explain and predict natural phenomena.

This view of natural kinds is avowedly pluralist, but it is less pluralist than Dupré's view, which he calls "promiscuous realism." I concur with him in thinking that different classification schemes reflect different interests and that there is no "uniquely best system of classification for all purposes or, which comes to the same thing, independent of any particular purpose" (1999, 473). However, unlike Dupré, I privilege epistemic purposes over other purposes and I therefore accord a special status to those classifications that are introduced primarily to serve those purposes. By contrast, Dupré argues that "Scientific classifications . . . are driven by specific, if often purely epistemic, purposes, and there is nothing fundamentally distinguishing such purposes from the more mundane rationales underlying folk classifications" (1999, 462). But, I would maintain, what distinguishes epistemic purposes from other purposes is that our best epistemic practices aim to uncover the divisions that exist in nature. Since the attempt to ascertain these divisions is none other than the search for natural kinds, classificatory schemes that fulfill epistemic purposes ought to be privileged over others in determining which categories are natural kinds. A category that serves, say, a purely aesthetic purpose cannot be expected to coincide with a natural kind. Consider the category *aquarium fish*, which applies to all and only fish that humans tend to keep in aquaria, largely for their aesthetic qualities. The fact that lionfish are thought to be desirable by fish enthusiasts while codfish are not, and that the former can be correctly classified as an *aquarium fish* while the latter cannot, is a fact about human aesthetic preferences. It does not mark a division between two kinds of fish, nor was it intended to do so. There are no generalizations to be made about aquarium fish (beyond the fact that they are all and only fish that are kept in aquaria by humans) and there is therefore no epistemic value to the category.[13] The same applies to the category *fish* when used in a loose rather than a precise scientific sense to pick out, roughly, all aquatic animals.

If natural kinds are classifications introduced for epistemic purposes, folk categories can be expected to correspond to natural kinds only when they serve an epistemic purpose. In these cases, they tend to be aligned with categories found in one or the other branches of the sciences or they become so aligned in the course of inquiry. When folk categories do not play an epistemic role, we should not expect them to correspond to natural kinds, and we should not expect the folk to defer to the experts. Unlike Quine, I do not think that folk categories will always be rejected in favor of scientific categories (though when they are not, they will tend to persist for nonepistemic reasons), and, unlike Dupré, I do not think that folk categories are *generally* as legitimate as scientific ones. In some cases, folk categories are revised or modified in such a way as to coincide with scientific categories (*consumption* and *tuberculosis*), in other cases folk categories drop out altogether (*hysteria*), and in yet other cases they remain in place to fulfill nonepistemic purposes (*fish*, in the sense of *aquatic animal*) and scientific categories are introduced alongside them.[14] It is only in the first type of case that we can expect our folk categories to correspond to natural kinds, since they (come to) coincide with categories that play an epistemic role. (Even some categories that pertain to some of the sciences may not serve an epistemic purpose; such claims have been made especially for some categories in the social sciences, and these claims will be examined in subsequent chapters.)

Some philosophers might react to the main proposal I am making by saying that it puts the epistemic cart before the metaphysical horse. But once we adopt a realist stance towards science, as I have done, we thereby

---

[13] It may be objected that facts about human aesthetic *preferences* are objective facts nonetheless, and may indeed be projectible. But if so, these facts would concern a different category – namely, *aquarium fish-keeper* – which picks out a kind of human being. But it is unlikely that this category would enter into the kind of inductive inferences and causal explanations that are characteristic of natural kinds. For more on the projectibility of human kinds, see Chapters 4 and 5.

[14] There is a further complication to this threefold classification of outcomes, which is nicely displayed by the examples cited. In the case of folk concepts adopted by science, sometimes the same term is retained but at other times a different term is introduced (as in the case of 'consumption' and 'tuberculosis'). Meanwhile, in cases in which a folk concept is retained alongside the scientific concept to serve some nonepistemic purpose, sometimes a different term is used, but at other times the same term is used and becomes ambiguous (as in the case of 'fish').

accept that the categories that science devises in order to understand nature provide the best insight into the kinds that really exist. The epistemic kinds that we arrive at as a result of the scientific enterprise comprise an effort to discern the nature of reality. It is not that epistemology is driving metaphysics, but that the epistemic enterprise of science attempts to reflect the divisions in nature, and those divisions mark the boundaries between natural kinds. In section 2.7, I will argue that this epistemic conception of natural kinds will lead us to a metaphysical theory of natural kinds, but first, in the following two sections, I will consider whether additional constraints should be imposed on scientific categories to weed some of them out as not corresponding to natural kinds.

2.4 FUZZY KINDS

I have argued in this chapter that natural kinds are epistemic kinds and that they will therefore correspond to the categories of science and not, in general, to folk categories. But this may be thought to be too pluralist or liberal an account of natural kinds, so it is worth considering some further conditions on natural kinds, to see whether natural kinds coincide with the entire range of categories proposed by the various branches of science or whether they are confined to some subset of scientific categories.

Mill places a further condition on natural kinds, in addition to those discussed in section 2.2 – namely, that there should be an "unfathomable chasm" (1843/1974, I vii §4) or "impassable barrier" (1843/1974, IV vii §4) between them. There is a certain metaphorical vagueness to these formulations, but I will take them to imply that the distinction between one natural kind and another is clear and that there are no intermediate instances between any two natural kinds. That is to say, when it comes to any natural kind, there is always a fact of the matter as to whether an individual does or does not belong to it. Let us call categories that satisfy this condition "discrete kinds" and those that fail to satisfy it "fuzzy kinds," and let us try to determine whether fuzzy kinds can be natural kinds, or whether an additional condition on natural kinds should be that they are discrete (not fuzzy).

To get a better idea of the content of this condition, it is worth distinguishing it from others with which it might be confused. Cluster or polythetic kinds, which were discussed in section 1.4, do not necessarily fail to satisfy this proposed condition on natural kinds. Indeed, even if a kind is structured as a *weighted* cluster, that does not preclude there being a precise cutoff between members and nonmembers. A simple example will serve to illustrate. Consider kind K, which is associated with properties $P_1, \ldots, P_7$ in such a way that each property is weighted according to the following factors, $P_1 = 0.5$, $P_2 = 0.4$, $P_3 = 0.4$, $P_4 = 0.3$, $P_5 = 0.2$, $P_6 = 0.1$, and $P_7 = 0.1$, and such that membership in K is dependent on achieving a score $s \geq 1.0$. In this case an individual possessing properties P_1, P_2, and P_6 definitely belongs to the kind while an individual possessing properties P_4, P_5, and P_7 definitely does not belong to the kind. In general, any individual that does not achieve the requisite score does not belong and there will be no intermediate cases. Despite the fact that cluster kinds are sometimes referred to as "fuzzy" kinds, they need not be structured in such a way as to allow intermediate cases.

There seem to be two ways in which a kind could have a fuzzy structure. The first type of fuzziness arises when a polythetic kind is associated with a cluster of properties in a vague or indeterminate way. I have said that polythetic and weighted cluster kinds need not be fuzzy kinds, since there may be a clear dividing line between members and nonmembers. But the above example of a weighted cluster kind could easily be modified to yield a fuzzy kind. Rather than a sharp cutoff, we could imagine a cluster kind with graded structure, whereby focal members of the kind achieve a score $s \geq 1.0$, more marginal members achieve a score such that $0.9 \leq s \leq 1.0$, and so on. In this case, there would be no requisite set of properties necessary and sufficient for membership in the kind and there would not even be a requisite set of property combinations. There would just be some property combinations that yield more or less clear members of the kind (and perhaps others that yield nonmembers). Another way in which a kind could be fuzzy rather than discrete would result from the kind being associated with one or more properties that vary continuously along some dimension. To keep it simple, consider the case in which one of the properties associated with the kind is dimensional. It may be that there is a range of values for that property that lead reliably to the instantiation of other properties (either singly or in conjunction with other properties), but that some values of the range do so more reliably than others or lead to the instantiation of more of those properties. In this case, we would also have graded membership in the kind, whereby some members are focal members of the kind and others are marginal members of the kind, or where some individuals are intermediate between that kind and a neighboring kind.

Is there a reason for imposing the discreteness condition on natural kinds? One of the main reasons for rejecting this condition is that it seems not to be satisfied by some paradigmatic examples of natural kinds – namely, biological species. Though it would be anachronistic to fault

Mill for failing to acknowledge this point, we now know from the theory of evolution that new species of organisms often arise as a result of gradual and incremental changes.[15] In the course of a process of speciation, there may well be individuals that are intermediate between two species. Hence, if this condition were to be imposed on natural kinds, it would preclude at least some (and perhaps most) species from being natural kinds. For rather different reasons, some chemical compounds may also have fuzzy boundaries, as shown by the example of isomeric compounds. Isomers are different substances composed of the same elements in the same proportions (e.g., ethanol, CH_3CH_2OH, and dimethyl ether, CH_3OCH_3). They are differentiated by their different structures, but, as Hendry (2006, 869–870) points out, molecular structure "is defined in terms of continuous variables like internuclear distances and angles between bonds." This means that the boundaries between isomeric chemical compounds are vague, and isomers are surely different chemical natural kinds since their chemical properties are generally very different. Thus, at least some chemical compounds are fuzzy kinds too, since they are associated with at least one property that varies continuously along some dimension.

Exclusion of some of the paradigmatic examples of natural kinds is a point against the discreteness condition, to be sure. But is there any positive reason to adopt it? One reason might be that a search for natural kinds is a search for the divisions that exist in the natural world. Hence, if we are to locate real divisions as opposed to inventing artificial ones, we should expect that they will be clearly marked in nature and that their boundaries will not be fuzzy or indistinct. In Mill's words, natural kinds should be separated by an "unfathomable chasm, instead of a mere

---

[15] Mill thinks that many distinctions between species, genera, and even families are not distinctions between true natural kinds; as he puts it: "Very few of the genera of plants, or even of the families, can be pronounced with certainty to be Kinds. The great distinctions of Vascular and Cellular, Dicotyledonous or Exogenous and Monocotyledonous or Endogenous plants, are perhaps differences of Kind; the lines of demarcation which divide those classes seem (though even on this I would not pronounce positively) to go through the whole nature of the plants. But the different species of a genus, or genera of a family, usually have in common only a limited number of characters. A Rose does not seem to differ from a Rubus, or the Umbelliferæ from the Ranunculaceæ, in much else than the characters botanically assigned to those genera or those families" (1843/1974, IV vii §4). Hence, the fact that many biological taxa fail to satisfy this condition would not seem to be particularly problematic from his perspective. But that view is not widely shared by other philosophers, and it was regarded as a definite drawback of his view by some of his earliest critics. Monck (1887, 638) wrote: "According to this doctrine [Darwinism] there are no such things as Natural Kinds separated from each other by impassable barriers; and whenever the line of demarcation between what I may call two adjacent kinds appears to be impassable, it is only because the intermediate members have perished in the struggle for existence." For similar criticisms, see also Towry (1887) and Franklin and Franklin (1888).

ordinary ditch with a visible bottom" (1843/1974, I vii §4). Taking Mill's metaphor a little further and drawing a loose analogy between the natural world and a natural landscape, we might distinguish between features such as deep ravines on the one hand and shallow trenches on the other. But Mill's analogy does not seem to vindicate the idea in question. Chasms are more prominent than ditches, to be sure, but if our concern is to discern all the features of a landscape, then we ought not to ignore the latter. They are no less real than the former. Divisions between kinds that are stark and well defined may be easier to discern than fuzzy boundaries, but that does not mean that the former are real divisions marking off natural kinds while the latter are not.

This point can be made in a less metaphorical fashion with the help of an illustrative example from phylogeny. Consider the following two hypothetical speciation events. In the first, there is a gradual and incremental shift from one stable form A_1 to another A_2, with many organisms intermediate between the two, while in the second there is a sharper transition between B_1 and B_2, which is due to one or more mutations that are responsible for giving rise to the new species. Once the two species A_2 and B_2 come into existence, does it matter that the boundaries between A_1 and A_2 are less sharp than the ones between B_1 and B_2, and does it mean that A_2 is not a natural kind while B_2 is? The mere fact that there are organisms in existence (or that were in existence at some point in the past) intermediate between A_1 and A_2 should not imply that A_1 and A_2 are not natural kinds.[16]

We have seen that there are grounds for thinking that two of the most widely acknowledged examples of natural kinds, chemical compounds and biological species, have fuzzy boundaries, and that realism towards natural kinds is not prima facie incompatible with their fuzziness. The case for fuzzy kinds can be made in a more positive fashion if we bear in mind the main characteristic of natural kinds as I have characterized them so far,

---

[16] It may seem as though there is a way of distinguishing species so as to erase the fuzziness of the boundaries between them. According to the phylogenetic species concept (PSC), two organisms belong to the same species if and only if they are capable of mating and producing fertile offspring. While this criterion for conspecificity has its limitations (e.g., it only applies to sexual species and is also not widely applicable to plants, which tend to be easily hybridized), it would seem to provide us with a more precise cutoff between two species, precluding the possibility of organisms intermediate between them. But even this criterion may not always lead to a clear result, since there are instances in which organisms may be anatomically capable of mating but do not, or are capable of doing so but only with difficulty, or can mate but only as a result of human intervention. There are also ring species, within which members of a population A can mate with members of population B, and B with C, but not A with C.

projectibility. If a kind *K* is associated with a set of properties that lead to the manifestation of yet other properties, then the relationships between these properties need not be strictly deterministic. Focal members may be ones that manifest more of the properties, or manifest them more frequently, or manifest them to a greater degree than nonfocal members. Furthermore, there may be no sharp distinction between members and nonmembers. Some cases may be ones in which there is continuous variation but a clear concentration around one or more foci, such as a bimodal distribution. Though there are intermediate instances, they may be relatively rare compared to the focal cases. Other cases may be ones of dimensional kinds, where not only is there continuous variation along some dimension *D*, but there is also a more or less continuous distribution of instances along the dimension in question.[17]

2.5 CROSSCUTTING KINDS

In the previous section, I argued that the naturalness of a kind *K* ought not to be revoked by its having fuzzy boundaries or by the existence of individuals that are intermediate between *K* and some other kind *K'*. A related issue has to do with whether the naturalness of a kind is affected by the fact that individuals belonging to that kind belong to some other kind as well. A traditional view about natural kinds has it that each individual can belong to one and only one natural kind. Natural kinds, as opposed to artificial categories, are such that they provide an exhaustive *and* mutually exclusive partitioning of the world of the individuals. Mutual exclusivity implies that for natural kinds $K_1, \ldots, K_n$, each individual can belong to only one of those kinds. This strong version of the *mutual exclusivity thesis* does not seem to have many adherents, but a slightly weakened version does. According to the weaker thesis, call it the *hierarchy thesis*, an individual can belong to more than one kind provided that the kinds it belongs to are arranged in a nested hierarchy. Any individual belonging to kind K_1 can also belong to other kinds on condition that those kinds constitute a series of increasingly inclusive kinds. Another way of putting this is by saying that natural kinds cannot *crosscut* or that there can be no *partial overlap* between natural kinds.

---

[17] In an illuminating discussion, Haslam (2002) draws a distinction between these two kinds of kinds, dubbing them "fuzzy kinds" and "pragmatic kinds," respectively. But the main difference concerns the mode of distribution of their members, so I regard them both as fuzzy kinds.

I have discussed the hierarchy thesis, and its denial, the *crosscutting thesis*, in other work (Khalidi 1993, 1998a), where I have provided various examples of systems of scientific categories that crosscut one another. In some cases, it might seem possible to dismiss one such system of categories (or both) as being illegitimate or not properly scientific. But there are many undeniably bona fide examples of scientific classification schemes that sort individual members into categories that do not coincide and are not wholly contained within one another. Moreover, there do not seem to be grounds in all such cases for ruling that one of the crosscutting schemes consists of nonnatural kinds. 'Nuclide' is the term for an atom type (not token) with a given number of protons and neutrons. Isotopic nuclides have the same number of protons; isobaric nuclides have the same total number of neutrons and protons. Nuclides can be categorized by atomic number (in which case, different *isotopes* of the same element will be classified together) or by mass number (in which case, different *isobars* of different elements will be classified together). Consider the nuclide *lithium-8* ($^{8}_{3}$Li), which can be classified either with other isotopes of *lithium*, such as *lithium-6* and *lithium-7*, or with isobars of other elements, such as *helium-8* and *beryllium-8*. In addition, if an atom of *lithium-8* comes to have more protons than electrons by losing one electron, it becomes a positively charged univalent *cation* Li^{+} and will share some properties with other monovalent cations as well as with cations in general (though not with other atoms of lithium). Although classification by atomic number and electric charge enables us to predict and explain its chemical properties better, classification by mass number is more efficacious for understanding its nuclear properties. Lithium-8 has different nuclear properties than do the other, more stable, isotopes of lithium, lithium-6 and lithium-7. It decays by electron emission (beta-minus decay) into beryllium-8 (which in turn decays into two alpha particles). Like helium-8, it has a short half-life and exhibits beta-minus decay.

Crosscutting taxonomic schemes are ubiquitous in the sciences. In biology, classification by species is the paramount mode of classifying organisms, but for many organisms classification by developmental stage is also efficacious in explaining features of those organisms. Many insects and amphibians undergo a larval phase, and developmental biology explains aspects of their morphology and behavior with reference to their being larvae. Many larvae are specialized for feeding, by contrast with mature adults who are more specialized for reproduction, and generalizations can be made about larvae that are related to this and similar properties. Hence, classification by developmental phase crosscuts classification by species (and higher taxa). In these cases and many others, two or more classification schemes sort individuals into systems of categories that are not nested wholly within one another but partially overlap. These crosscutting systems are not rivals that belong to competing scientific theories, but can coexist comfortably because they pertain to different interests (Khalidi 1998a). Though the categories are bona fide scientific ones and the interests they serve are broadly epistemic, the specific explanatory interests served by different categories diverge somewhat within different scientific disciplines and subdisciplines, such as the need to explain chemical properties as opposed to nuclear properties or the need to explain developmental properties across species as opposed to the need to explain properties of organisms within a species. (Recall Mill's point from section 2.2 that different branches of science may regard different properties as being important.)

Given the preponderance of evidence against it, what is the motivation for affirming the mutual exclusivity thesis? The thesis is rooted in a long philosophical tradition of theorizing about natural kinds that regards them as uniquely classifying the natural world into a set of categories, and this view is closely associated with essentialism.[18] If each individual belongs to exactly one natural kind, and each natural kind is associated with an essence, then each individual has a unique essence that makes it what it is and not another kind of thing. Hence, for each individual there is one correct answer to the question, what is it? The answer is that it is a member of kind K, whose essence consists of a property or set of properties. The picture of the universe that this leads to is undeniably neat, each individual belonging to just one natural kind. But the attraction is specious. As I mentioned above, hardly anyone advocates the strong thesis of mutual exclusivity, and even essentialists allow hierarchies of nested kinds. The weakened hierarchy thesis allows an individual to belong to a series of kinds as long as those kinds are all contained within one another. But this weakening of the thesis already drains away much of its apparent appeal. If each individual is allowed to belong to a series of kinds, then we can no longer say that every individual in the universe belongs to some unique kind that supplies its distinctive essence. On the weakened version of the thesis, each individual belongs to a series of more inclusive kinds. But if an individual atom is allowed to belong both to the kind *lithium-8* and to the more inclusive kind *lithium*, then why not allow it also, if it happens to be

---

[18] Moreover, I have argued that the hierarchy or uniqueness thesis is needed as a premise in one influential argument for contemporary essentialism (Khalidi 1993).

ionized, to be a member of the kind *cation*? The category *cation* includes
those positively charged atoms that have more protons than electrons, a
property which in turn accounts for many of their chemical properties.
A positive lithium ion has the property of being able to burn in oxygen to
form lithium oxide (Li_2O). This is a property it shares with other positive
ions such as the magnesium ion, which burns in oxygen to form
magnesium oxide (MgO), though it does not share this property with
atoms of lithium that are not ionized. Similarly, if an individual organism
classified as a *monarch butterfly* (*Danaus plexippus*) can belong to the more
inclusive kinds *insect* and *Lepidoptera*, why not allow it to belong to the
kind *larva*, all members of which share certain important behavioral
properties such as efficiency at locating food sources? In addition to the
hierarchy of nested categories to which these individuals belong, they also
belong to a set of categories that crosscut those others, overlapping with
them only partially. If belonging to a multiplicity of kinds that is hierarchically arranged is allowable, then it is not clear why belonging to a
nonhierarchical multiplicity of kinds should be disallowed.

The picture of natural kinds that emerges as a result of the denial of the
hierarchy thesis is an avowedly pluralist one, according to which there are a
large number of crosscutting kinds, some more natural than others
depending on the epistemic access that they afford to certain aspects of
reality (this will be further justified in section 3.6). Natural kinds correspond to scientific categories introduced in order to enable us to make
inductive generalizations and to explain naturally occurring phenomena.
But while pluralist, this account of natural kinds is less pluralist than
Dupré's account, since he seems to admit even nonepistemic kinds into
his ontology, as I explained in section 2.3.

2.6 HPC KINDS AND CAUSAL KINDS

I have stressed that genuine natural kinds ought to be associated with a set
of scientifically important properties, but have not said enough about how
those properties are linked. That is the ingredient supplied by one of the
most prominent contemporary accounts of natural kinds – namely, the
"homeostatic property cluster" (HPC) account of natural kinds advocated
by Richard Boyd (1989, 1991, 1999a, 1999b). On the face of it, the HPC
account of natural kinds is inimical to the essentialist view that kinds are
associated with a set of necessary and sufficient properties, which was
criticized in the previous chapter (though I will also mention some
attempts to square it with essentialism later in this section). It allows for

the existence of polythetic or cluster kinds, contrary to the standard understanding of essentialism. Moreover, according to Boyd, it is not enough for there to be a (loose) cluster of properties associated with a natural kind; those properties must be so associated for a reason: They are kept in equilibrium by a causal mechanism. Boyd's HPC account of natural kinds states that every kind is associated with a set of properties not by happenstance but because there is some "underlying mechanism" that gives rise to all of them or because the presence of some of them favors the presence of others. Properties, $P_1, \ldots, P_n$, associated with a kind K are "contingently clustered" in nature, and this is not a cosmic coincidence but is rather the result of a process of "homeostasis" as a result of which these properties are kept in equilibrium. Boyd also recognizes that the properties associated with a kind need not be possessed by every member of that kind, and he calls this "imperfect homeostasis." In such cases, only some of the homeostatic mechanisms might be present that hold such properties together. Moreover, the properties associated with a kind may vary over time since there is no single property (or subset of properties) that is necessary for membership in the kind (Boyd 1989, 16–17; cf. Boyd 1991, 143–144).

Biological species are widely thought to be a good fit for the HPC account of natural kinds. The HPC account clearly accommodates the fact that there is no set of genotypic or phenotypic properties that is both necessary and sufficient for belonging to a species, as most biologists and philosophers of biology now believe. The account also holds that the properties associated with each natural kind are held together as a result of a causal mechanism or set of mechanisms. In the case of biological species, the principal mechanism is *interbreeding* according to Boyd, which ensures that properties possessed by members remain in circulation within the population. Others have added mechanisms of *genetic descent* and *environmental pressures* to the mix, on the grounds that there are multiple causes that hold a biological species in homeostasis in addition to inter-breeding among members of that species (Wilson *et al.* 2009). Finally, the HPC account also makes room for the evolution of species, allowing that the properties associated with a kind can change, so long as there are mechanisms holding the kind in a state of equilibrium.

The HPC account of natural kinds would seem to posit something that any account of natural kinds should – namely, causal relations among the properties associated with a natural kind. In fact, it posits the existence of a *causal mechanism* that holds together the properties associated with a kind. As Wilson *et al.* (2009, 199) put it, the mechanism ensures that these properties constitute a *cluster* rather than a mere *set*. Moreover, as they also explain, once the existence of the properties within the cluster is understood to spring from certain causal mechanisms, this assures us that the properties have not been associated with each other on artificial grounds, merely as a result of our predilections to group certain properties together (Wilson *et al.* 2009, 198). The HPC account also has some additional benefits, which apply broadly to other cluster kinds, not just biological species. With respect to cluster kinds, the account provides a principled explanation for why some individuals should be considered members of the kind and others not. In considering "weighted cluster" kinds in section 2.4, I presupposed that they were such that membership was determined by attaining some "score" depending on the properties possessed and the weighting of each property. But might cluster kinds be structured in such a way that there is no such formula determining membership? On the HPC account, this is allowed because the mere possession of properties is not the whole story. In the case of HPC kinds, the causal mechanism or mechanisms that hold those properties in place is crucial to the account and serve to determine whether an individual belongs to a kind. Individual organisms may lack some of the properties associated with a biological species, yet they may belong to the species nonetheless, since they are subject to the very same mechanisms that have led to the instantiation of those properties in other members of the species, including interbreeding, genetic descent, or environmental pressures.

The HPC account of kinds has a number of strengths, not least because it lends greater credence to the viability of cluster or polythetic kinds as natural kinds. But it also has some shortcomings. It is not that the HPC account of natural kinds never applies to natural kinds; it often does. However, I would argue that it would be a mistake to conclude that all natural kinds are HPC kinds. Typically, proponents of the view claim that the account pertains primarily to biology rather than to physics or chemistry, so it is not even meant as a complete account of natural kinds. But even in biology, the HPC account need not apply to a category for it to qualify as a natural kind. Although it is a useful framework for understanding why some kinds are natural kinds and is a convenient reminder of the causal dimension of natural kinds, it does not seem to fit many apparently natural kinds.

The application of the HPC account of natural kinds to biological species has been challenged by some philosophers of biology, who find that it prioritizes similiarity among members of a species as a criterion for species membership over descent from a common ancestor. Even though one of the homeostatic mechanisms cited by HPC theorists when it comes to species is genealogical descent, Ereshefsky (2010) thinks that the HPC account implies that what makes a species a kind is the similarity among its members, as opposed to descent or commonality of origin, whereas most biological systematists emphasize the latter.[19] When descent and similarity diverge, biological systematics chooses descent, whereas the HPC account opts for similarity, according to him.

The root of the problem is that HPC theory assumes that all scientific classification should capture similarity clusters. However, that is not the aim of biological taxonomy. Its aim is to capture history. (Ereshefsky 2010, 676)

Defenders of the HPC account may respond by saying that this is true primarily of the cladistic approach to taxonomy and that other approaches to taxonomy also factor in other properties when classifying biological species. Cladists consider that speciation has occurred if and only if there has been branching in the phylogenetic tree (cladogenesis), whereas some other systematists assess speciation on the basis of gradual divergence of traits (anagenesis). On the latter view, speciation may occur without branching provided enough genetic mutations have occurred. In these cases, the cluster of genetic properties is considered more salient than the mechanism of evolutionary branching, according to some noncladistic systematists. Hence, it does not seem that the objection is fatal to the attempt to apply the HPC account to biological species, at least if one is not a strict cladist about taxonomy. Be that as it may, problems also afflict the HPC account when it comes to higher biological taxa, such as genera, families, classes, and so on. Here, the only serious candidate for a mechanism is genealogical descent (since interbreeding is out of the question and members of higher taxa are not generally subject to the same environmental pressures). But if that is the case, then it might seem as though there is no work left to do for the homeostatic property cluster. The kind is instead equated with a certain lineage in the phylogenetic tree, and any shared traits that exist among members, if indeed they do exist, are mere byproducts of that common evolutionary history.

At this point, a natural modification of the HPC account might suggest itself – namely, one in which a kind is identified with the mechanism or

---

[19] Ereshefsky and Matthen (2005) also criticize HPC on the point that there is widespread dissimilarity among members of a species, and that this is not just accidental but is also central to any biological account of species. Indeed, some of the causal mechanisms in question are heterostatic in the sense that their job is to maintain variation in the population (e.g., dimorphism or polymorphism). Wilson *et al.* (2009) respond to some of their points.

mechanisms that keep it in homeostasis. As I mentioned above, some proponents of the account argue that the causal mechanisms when it comes to biological species can include genetic descent, as well as interbreeding, and environmental pressures. Since the mechanisms are supposed to be responsible for giving rise to the properties possessed by members, perhaps those mechanisms correspond to "deeper" or "underlying" properties that generate all the other properties. This may allow it to evade some of the criticisms raised against it as an account of biological species, since the mechanisms of interbreeding, genetic descent, and environmental pressures might be used to individuate species, though they would have to be different for different species in order to distinguish them from one another. As Boyd (1999a, 170) observes, when it comes to species in particular, "the homeostatic mechanisms important to the integrity of a species vary from species to species." Some philosophers have elaborated the HPC account along these lines when it comes to biological species, regarding the homeostatic mechanism as being essential to a biological species. Indeed, a curiosity of the HPC account of natural kinds is that some of its advocates consider it to be compatible with essentialism – indeed, to be a form of essentialism – while others regard it as an alternative to essentialism. Griffiths (1997, 212) writes: "The essence of a kind is its causal homeostatic mechanism – whatever it is that explains the projectibility of that category." He states that in equating essences with "causal homeostatic mechanisms," he is following Boyd (1991b, 1999b). But Boyd himself does not think of these mechanisms in terms of the standard specification of essentialism that I have tried to explicate in Chapter 1.[20] Boyd (1999b, 169) writes: "The natural kinds that have unchanging definitions in terms of intrinsic necessary and sufficient conditions . . . are an unrepresentative minority of natural kinds (perhaps even a minority of zero)."

Whether or not we understand it in essentialist terms, if the HPC account is modified in such a way that the mechanism rather than the cluster of properties is taken to individuate the kind in question, we run into a different problem, at least if it is to be a general account of natural kinds (cf. Craver 2009). Even if the mechanism is not considered to be the "essence" of a natural kind, the HPC account gives pride of place to the

---

[20] Notably, Griffiths (2002, 72) has excoriated "folk essentialism" in biology as follows: "Folk essentialism understands biological species as the manifestation of underlying 'natures' shared by all members of a species. . . Since folk essentialism is both false and fundamentally inconsistent with the Darwinian view of species, it should be rejected." But what Griffiths objects to is not essentialism *per se* but a particular brand of it.

mechanism and assigns the property cluster a secondary role. But in many cases, the relationship between mechanisms and properties is not nearly so neat. There are biological kinds for which a homeostatic mechanism seems crucial and, as it were, holds the kind together (see section 5.5 for one example). But there are other kinds, even other biological kinds, for which there may be no single, well-defined mechanism, or for which some of the properties associated with the kind cause others, or for which there is a self-sustaining process at work, as when properties present at one stage of development give rise to properties at another stage of development, which in turn give rise to the former properties in the next generation. This last type of relationship need not involve a metaphysically suspect type of "self-causation," but just the familiar efficient causation operating across successive life cycles. Consider the process at work in maintaining the properties associated with the kind *larva*, mentioned in the previous section. The larva's adeptness at finding food is what (partly) causes the emergence of a mature imago, whose success at reproduction is what gives rise to the next generation's larvae, which in turn have traits designed for locating sources of food, and so on. Here, we do not seem to have a central causal mechanism that is responsible for a host of properties, but rather a set of self-sustaining causal property instances.

It is not even clear that homeostasis is strictly necessary for the existence of a kind. Most species evolve, and the properties associated with them are not maintained in a strict state of equilibrium. As a result of mutation and natural selection, some of the properties possessed by members of a species are lost and others acquired, so there is a constant process whereby the properties associated with a species are altered (cf. Ereshefsky and Matthen 2005). Sometimes this leads to speciation and the emergence of a new kind altogether, but often the same species persists despite considerable divergence, and there is theoretically no upper limit on the extent to which members of a species might diverge from an ancestral form. The problem with a homeostatic account of species is that it seems to presuppose that there is some ideal or normal state that is being maintained by causal processes. But modern biology has disabused us of the notion of a "natural state model" of species, according to which "variability within nature is . . . to be accounted for as a deviation from what is natural" (Sober 1980, 360). On that kind of outdated typological thinking, there is some natural type to which all specimens tend to converge, and all specimens that do not conform to this type are deviations from the norm. This model has been rendered obsolete by one that regards variability among members of a species as being the norm itself rather than divergence from the norm.

Positing a homeostatic mechanism in each species that tends to keep the properties in equilibrium is at odds with this way of thinking about species. Therefore, even when it comes to species, the paradigmatic biological kind, there are strong grounds for thinking that the HPC account is not a good fit.

Why not, then, give up on the idea that homeostatic mechanisms are centrally important to natural kinds? As Craver (2009, 579; original emphasis) speculates:

It is possible . . . to reject [the homeostatic mechanism] and to keep the rest as a *simple causal theory* of natural kinds. According to this view, natural kinds are the kinds appearing in generalizations that correctly describe the causal structure of the world regardless of whether a mechanism explains the clustering of properties definitive of the kind.

To this, I would add that the account of natural kinds as epistemic kinds already incorporates the causal component of the HPC, since inductive generalizations in science are ultimately underwritten by causal relations. Boyd (1991, 139) himself makes this point well:

Kinds useful for induction or explanation must always "cut the world at its joints" in this sense: successful induction and explanation always require that we accommodate our categories to the causal structure of the world.

But the causal relations will be more variegated and diverse than the HPC account seems to allow. In some cases, the mechanism is separate from and is the common cause for the properties associated with the kind. In other cases, the mechanisms involved may be incorporated into the set of properties. In yet other cases, there may be nothing that deserves to be called a "mechanism" at all. In at least some of his formulations of the account, Boyd allows that when it comes to some natural kinds, the presence of some properties favors the presence of others, thus seeming to renege on the need for a homeostatic causal mechanism in all cases.[21] But if so, then the existence of a homeostatic mechanism is incidental and ought not to be the guiding principle of the account. In this vein, I am arguing that mechanisms need not be involved at all and that when they are, they need not be the cause of all the properties associated with the kind and need not keep all the properties in homeostasis.

---

[21] "Either the presence of some of the properties in [a family of properties] *F* tends (under appropriate conditions) to favor the presence of the others, or there are underlying mechanisms or processes which tend to maintain the presence of the properties in *F*, or both" (Boyd 1989, 16).

The HPC account of natural kinds rightly draws our attention to the fact that there is a causal connection between some of the properties associated with a natural kind and others. If natural kinds are to play a role in inductive inference and serve the purposes of science, then they will be implicated in causal processes. At an early stage of inquiry, scientists interested in projecting from some properties to others will typically observe or seek correlations between properties or sets of properties. Upon closer inspection, they will discern causal relationships among these properties or their instances. They find that when a (possibly loose) cluster of properties, $P_1, \ldots, P_m$, is instantiated, it tends to give rise causally to another (possibly loose) cluster of properties, $P_{m+1}, \ldots, P_n$, which may in turn give rise to others. Alternatively, the second cluster may tend to cause another iteration of the instantiation of the first cluster, in a cyclical fashion. There may also be more intricate causal relations among members of these clusters of properties. We then identify natural kinds either with the first subset of properties or with the entire set. Sometimes, the first subset of properties, which are causally prior, can be considered the set of "primary" properties of the natural kind, while the second subset can be considered "secondary" (see section 6.2). However, that does not mean that there will always be some causal *mechanism* that holds the properties in the cluster together, or that those properties are always held together in a state of homeostasis.[22]

2.7 CONCLUSION

In this chapter, I began by defending an epistemic conception of natural kinds, according to which natural kinds are picked out by those categories that appear in our completed scientific theories. They are captured by the categories that best enable us to organize our knowledge of nature, are projectible and figure in inductive generalizations, successfully explain the phenomena, and so on. But this characterization of natural kinds can be accused of having a "Euthyphro problem." In an echo of Socrates' response to Euthyphro on the subject of piety, it could be objected that a kind is not natural because it is a scientific kind; rather it is a scientific kind because it is natural. In other words, the fact that a kind plays a role in science is not an intrinsic fact about that kind itself and is surely not what

---

[22] See Slater (2009) for similar criticisms of the HPC theory of natural kinds as applied to proteins; Slater (2010) moves in the direction of a more liberal "stable property cluster" theory of natural kinds.

makes it natural. There is something to this objection and it points to the fact that discoverability by science is not the whole story about natural kinds. Science aims to identify projectible properties, particularly clusters of properties that point reliably to yet other property clusters. The fact that these properties are projectibly clustered indicates that there are causal links between them. Hence projectibility is the epistemic marker for the metaphysical relation of causality. But projectibility does not indicate causal priority or specify the direction of causation. If property P_1 is projectibly linked to P_2, it may be that P_1 causes P_2, or that P_2 causes P_1, or that P_1 and P_2 are effects of a common cause P_3, or that P_2 causes P_4, which causes P_1, and so on. Scientific inquiry has the means to distinguish these different possibilities and to identify those properties that, when coinstantiated, tend to cause the instantiation of other properties. Such causal properties or clusters of properties are associated with natural kinds. What started as an epistemic conception of natural kinds has led us to a characterization of natural kinds in terms of causality, or, in other words, in metaphysical terms.

The projectibility of natural kinds, their role in inductive inference, and their explanatory and predictive value reflect the causal relationships in which they participate. But there is no single causal template that fits all instances of natural kinds or relates natural kinds to their associated properties. Philosophers from Mill to Boyd are right to emphasize the importance of causal relations to natural kinds, since causality is what holds together the properties associated with natural kinds. But I took issue with Boyd's idea that there is a causal mechanism that maintains all these properties in a state of equilibrium, since the causal story seems more complicated for many natural kinds (this claim will be further corroborated in Chapters 5 and 6). At a minimum, philosophical naturalism enjoins us to take seriously a posteriori evidence from science and this leads us inexorably in the direction of what Craver (2009) calls a "simple causal theory" of natural kinds. This theory seems capable of accounting for a wider range of natural kinds in the sciences, and I will try to elaborate on it further in the following chapters as further case studies of scientific kinds are examined.

On a simple causal theory of natural kinds, a natural kind is associated with a set of properties whose co-instantiation causes the instantiation of other properties. This makes room for the possibility that the causal efficacy of natural kinds may be a matter of degree on a number of different dimensions. Some clusters of properties give rise to a rich network of other properties while others only give rise to a few. Some properties are linked by deterministic causal relations while others are linked by weaker causal relations. Some properties give rise to others under a wide range of background conditions while others only do so in a narrower set of circumstances. Some are instantiated relatively rarely in the universe while others are ubiquitous. All of these differences make for differences in the natural kinds involved, and they may render the naturalness of a kind a matter of degree. Nevertheless, whether or not a set of properties is causally responsible for the instantiation of other properties is an objective matter, even if it admits of degrees (I will revisit this issue in section 6.3). If we take our cue from Mill and compare the natural world again to a landscape (as in section 2.4), some features of the landscape may be more prominent than others, and some may dominate others due to height or width or gradient, but that is not to say that the existence of any of these features is subjective.

CHAPTER 3

Kinds in the special sciences

3.1 WHICH SCIENCES ARE SPECIAL?

In the previous chapter, I argued for a naturalist conception of natural kinds, according to which natural kinds are those kinds that are discoverable by science and feature in our considered scientific theories, and this led me to defend a "simple causal theory" of natural kinds. But philosophers have tended to think that not all sciences are created equal, dividing them into the basic sciences and the nonbasic, or "special," sciences. Since the categories posited by the latter have not been considered natural kinds by many philosophers, this chapter will look more specifically at kinds in the special sciences. The task is complicated by the fact that there is no firm philosophical consensus on which sciences these are. It is fairly uncontroversial to say that physics is *not* included among the special sciences, and perhaps not chemistry. But even the exclusion of physics from the special sciences is not unproblematic, since such branches of physics as solid-state physics, geophysics, fluid mechanics, and biophysics should surely be considered special sciences (this claim will be justified further in due course). Similarly, those who would include chemistry among the basic sciences may rightly want to exclude biochemistry, ecological chemistry, and perhaps even organic chemistry. Hence, it is not a simple matter to say that physics and chemistry fall along one side of the divide and all the other sciences fall on the other. But it does seem safe to say that the basic sciences include, at the very least, elementary particle physics, while the special sciences include such branches of science as biology, geology, meteorology, zoology, and neuroscience, as well as social sciences such as sociology and psychiatry. In Chapter 1, I argued against the view that the only natural kinds are microphysical kinds, providing reasons for thinking that elementary particles and their properties are not the only real kinds and properties. But even if one accepts that the dividing line between natural and nonnatural kinds should not be drawn at the level of elementary particles, it is not obvious how far to venture beyond the microphysical domain. Are there good grounds for restricting natural kinds to some sciences and not others? And what makes such sciences capable of harboring natural kinds while other sciences are not? The aim of this chapter is to determine whether there are any grounds to disqualify special science categories from corresponding to natural kinds.

I will begin by criticizing some very general arguments for discounting the existence of kinds and properties in the special sciences, which claim that all "higher-level" kinds and properties either are mere disjunctions of "lower-level" kinds and properties or are reducible to them. Then, I will consider a closely related argument, which holds that special-science kinds and properties cannot have causal efficacy since all the causal work must be done at a "lower level." In addition, it is sometimes argued that there are no laws in the special sciences or, if there are, they are very different in character from the laws of the nonspecial sciences. I will try to cast doubt on this view, further erasing the distinction between special-science and basic-science properties and kinds. These arguments will bolster the case for a "simple causal theory" of natural kinds introduced in Chapter 2. But that theory is challenged by the observation that causal patterns are ubiquitous in nature and that natural kinds will be too numerous and ineffectual to be worth the name. I will try to respond to that challenge and will further defend the idea that systems of natural kinds can crosscut one another because they pertain to different aspects of the natural world.

3.2 MULTIPLE REALIZABILITY AND SPECIAL-SCIENCE KINDS

Should the natural kinds include among their number the kinds of the special sciences? I argued in Chapter 1 that the properties associated with natural kinds should not always, despite essentialist claims, be microphysical, and put forward some considerations to support this claim. But it may still be thought that natural kinds should be restricted to the basic sciences – namely, basic physics and chemistry. One reason for thinking so is the idea that the categories deployed in the special sciences always correspond *ultimately* to more basic kinds. Though opponents of special-science kinds may allow that they are multiply realizable in more basic kinds, they may resist the claim that they are natural kinds in their own right because every specific instance of a special-science kind corresponds to an instance of a more basic kind (though perhaps not the same basic-science kind on every occasion or in every system), and it is this more basic kind, not the special-science kind, that ought to be admitted into our ontology. Against this view, there is a well-known argument for special-science kinds due to Fodor (1974, 1997), which has been widely accepted by some philosophers. But the argument has recently come in for some sustained criticism by philosophers who have denied that the multiple realizability of special-science properties and kinds guarantees their autonomy from the basic sciences (Kim 1992, 1998, 2003; Papineau 2009; Shapiro 2000). Instead, these philosophers have argued that these properties and kinds are either reducible to those of the basic sciences or are not genuine kinds but mere disjunctions of more basic kinds.

To address the issue, I will be taking a look at some of the properties and kinds involved in the special science of fluid mechanics. Though some might think that this is a basic science by virtue of being a subbranch of physics, there are two considerations that I think are decisive in deeming it to be a special science. The first is that it is a macrolevel science at least some of whose properties and kinds simply have no counterparts at the microlevel (e.g., the property *viscosity* and the kind *Newtonian fluid*) and are not properties and kinds of atoms and molecules (much less elementary particles).[1] The second is that this science manifests the distinctive aspect of the special sciences – namely, multiple realizability and substrate neutrality. The properties and kinds of fluid mechanics can be realized very differently in different media, say, liquids and gases, as will be seen shortly.[2] There are a number of important reasons for looking at examples from these sciences rather than the more familiar cases described by philosophers, such as biology, psychology, and the social sciences. Biology has been considered by many philosophers to be a special case by virtue of the mechanism of natural selection and the presence of design, while psychology and the social sciences are marked by intentionality. The choice of a special science that is nonbiological (much less psychological or social) makes some of the phenomena surrounding special-science properties and kinds more tractable. Later, the possibility of natural kinds in the biological and social sciences will be examined in Chapter 4.

The phenomena that I will be exploring in this section pertain to the macroscopic behavior of fluids, specifically some of the principles governing their flow and their diffusion through other fluids. Let us start

---

[1] Compare Ross and Spurrett (2004, 604): "By 'special' science we have in mind any science not concerned with justifying, testing, or extending the generalizations of fundamental physics, and hence most science, including ... most of physics."

[2] Another common way of characterizing the special sciences is by saying that they apply to specific types of systems or pertain to particular regions of the universe rather than the universe as a whole. But I think this characterization is problematic for reasons to be elaborated in section 3.4.

by considering Fick's first law of diffusion, which describes the diffusion of one fluid into another; this law relates the rate of fluid diffusion to concentration of the fluid, distance traveled, and a diffusion coefficient. In symbols, it can be expressed as:

$$J = -D(\delta C / \delta x)$$

where J is the flux or rate of diffusion, D is a diffusion coefficient, C is concentration, and x is distance traveled (the negative sign indicates that flux is positive when diffusion goes from high to low concentration, i.e., in a negative direction). In other words, the flux or diffusion rate is the product of some coefficient (which varies with the fluid involved) and the change in concentration over distance. Intuitively, it can be interpreted to say (in part) that when two fluids with different concentrations come into contact, fluid flows from an area of greater concentration to an area of lesser concentration, and the steeper the concentration gradient the higher the rate of diffusion. The law has been found to apply not just to liquids and gases, but also to the diffusion of fluids through some porous solids, and other systems. The diffusion coefficient (D) cited in the law depends on *viscosity* (as well as on temperature and other factors). The relation between the diffusion coefficient and viscosity is given by the Stokes–Einstein equation:

$$D = k_B T / 6\pi\eta r$$

where k_B is Boltzmann's constant, T is absolute temperature, η is viscosity, and r is the radius of the diffusing particles. *Viscosity* is a measure of the resistance to flow within a substance, which is a good candidate for a special-science property that is multiply realized in different contexts. In liquids, *viscosity* is strongly associated with the strength of the chemical bonds between molecules in the liquid, which influences the rate of diffusion through that liquid and the rate of flow of that liquid. To a first approximation, viscous liquids flow more freely, while nonviscous ones are sluggish; this is in large measure a reflection of the nature and strength of their chemical bonds. In gases, *viscosity* is a function of physical quantities such as the density of the gas and the temperature, since gas molecules are relatively far apart and are not bound together like the molecules of a liquid. In porous solids, such as soils, *viscosity* is largely dependent on the dimensions of the pores relative to the fluid particles that are diffusing through the solid. Since Fick's first law can be shown to apply in a wide range of cases, and since *viscosity* can take on a number of different manifestations in diverse domains, including gases, liquids, and even solids, the property of *viscosity* is multiply realized relative to properties in the basic sciences. Indeed, this property would seem to exemplify the multiple realizability and substrate neutrality of properties in the special sciences. This appears to vindicate the claim that there are special-science properties, such as *viscosity*, that feature in laws and generalizations in the special sciences and are multiply realized in microphysical basic-science properties, in the sense that they are dependent on different basic-science properties in different domains (e.g., strength of chemical bonds between molecules in liquids, density of molecules in gases, and size of pores in solids). Moreover, the special-science property of *viscosity* is also associated with special-science *kinds*. For example, a *Newtonian fluid* is a kind of fluid that has *constant viscosity* (for a given pressure and temperature); that is, the viscosity does not depend on the magnitude of the force acting upon the fluid. By contrast, a *non-Newtonian fluid* is one whose viscosity depends on the magnitude of the force applied to the fluid. For example, a *Newtonian fluid* flows the same no matter how hard you stir it, whereas a *non-Newtonian fluid* may flow more the harder you stir (e.g., nondrip paint thins out), or flow less the harder you stir it (e.g., pudding stiffens). In some non-Newtonian fluids, the *viscosity* can even depend on the rate of the applied force not just its magnitude (e.g., the children's toy "Silly Putty," which flows if you pull it slowly but snaps if you tug on it quickly).

Some philosophers are not convinced by a consideration of such cases and insist that special-science properties such as *viscosity* ought to be reducible to the basic-science kinds that underwrite them. One way of understanding what higher-level "properties" are for these philosophers is to see them as functional properties (though, as we shall see, it is misleading to consider them to be real properties), which specify certain functional *roles* that have different *occupants* or *realizers* in different types of systems. *Viscosity* may seem to fit this account very well, since although it can be characterized functionally in terms of its role, roughly the resistance to flow, the occupant of this role can vary from substance to substance, or system to system. To simplify somewhat for the sake of argument, the same causal role or function that is played by the strength of the chemical bonds in a liquid can be played by the density of the molecules in a gas, or the size of the pores in a porous solid. Even though there is no single basic-science property that corresponds to *viscosity* in the diverse situations in which it is manifested, it can be given a functional characterization that is in turn reducible to a basic-science property in each substance or type of substance. Such a "functional reduction" (Lewis 1972/1999) is different from the traditional notion of an across-the-board reduction of a special-science property to a basic-science property, which would require a uniform or "type-type" equivalence between the two sets of properties, macro- and micro-. Yet, as we shall see, it seems to lead inexorably to the conclusion that there are no genuine special-science properties, since the real occupant of the role specified by these higher-level properties is a basic-science property.

According to this proposal, any general reference to *viscosity* in natural laws such as Fick's first law would be replaced by the property that, in some particular system or in some set of circumstances, underlies the fact that the substance in question is resistant to flow (to a certain degree). In gases it would presumably be a physical property that depends primarily on the density and temperature of the molecules, while in liquids it would be a chemical property that depends on the nature and strength of the chemical bonds between the molecules. If we were to replace *viscosity* in each of these cases with a term or terms pertaining to the microphysical realm, we would be left with a number of more specific laws that describe flow in these different contexts, with a different term to stand in for the diffusion coefficient, and hence *viscosity*, in each of these cases. Fick's first law might then be considered the disjunction of all these other more specific laws. But it is widely agreed among philosophers on both sides of this debate that the disjunction of laws is not generally a law, the disjunction of properties is not generally a property, and the disjunction of kinds not generally a natural kind (as I already argued in Chapter 1). Hence, the inevitable conclusion of this line of argument is to repudiate the kinds and properties that appear in the special sciences, as well as the laws and generalizations that feature them. Perhaps "Fick's first law" should not have been deemed a law, contrary to our initial assumption (and standard scientific presentations of this topic), and properties such as *viscosity* should not be considered real properties. This is why many philosophers draw a distinction between functional properties and kinds and microstructural properties and kinds, denying that the former can be real properties and natural kinds. In this vein, Kim (1992) has argued that the fact that purported special-science properties and kinds correspond to disjunctions of basic science properties and kinds shows that the former are not real properties and natural kinds at all (and that the generalizations in which they appear are not genuine laws). He compares such alleged kinds with the category *jade*, which picks out all and only instances of two different minerals with distinct chemical structures, *jadeite* and *nephrite*. Each of these minerals is a kind, to be sure, but *jade* is not a kind in its own right, since the only thing that samples of jade have in common is that they are samples of a shiny, green mineral. In a similar fashion, it might be said that special science properties, like *viscosity*, are not genuine properties, special-science kinds like *Newtonian fluid* are not natural kinds, and special science laws, like Fick's first law, are not genuine laws.

A related challenge to special-science properties and kinds can be put in terms of a dilemma, which has been clearly articulated by Shapiro (2000). Consider some special-science kind *K*, which is allegedly multiply realizable and can be characterized in functional terms. What would make such a kind multiply realizable is that at least some different instances of *K* have genuinely different properties. However, to avoid trivializing the notion of multiple realization, we need to further specify that the different properties should be ones that are causally relevant to fulfilling the function in question, or that the different instances fulfill the function in genuinely different ways. (Otherwise, Shapiro claims, differently colored corkscrews would count as multiple realizations of the purported kind *corkscrew*.) But now the dilemma looms large. If the different instances of *K* have genuinely different (causally relevant) properties, then they should surely be counted as instances of different kinds, and there is no multiple realization of a single kind *K*, but *multiple kinds*, say, *L*, *M*, *N*, and so on. But if they do not have genuinely different causal properties, then they should be considered members of the *same kind*, and there is no multiple realization after all. In the first case, we have a functional category *K* that picks out a role or function but does not enter into any causal laws in its own right. In the second case, we have a single realization in terms of genuine causal properties, but these will pertain to more basic sciences. Interestingly, in this latter connection, Shapiro (2000, 650) touches on the property of *viscosity*, which has served as my primary example:

There exist laws about the viscosity, specific gravity, freezing point, and so on of all samples of water, because all samples of water are composed of H_2O and are thereby determined, according to the microlevel laws that describe the behavior of H_2O, to exhibit similarities in their viscosity, and so on.

Hence, either special-science properties and kinds are reducible to microphysical ones, or they pick out certain analytically defined functions that do not enter into genuine causal laws. In the former case, as in the case of laws about the viscosity of water, the properties and kinds concern specific substances with specific microstructures. In the latter case, Shapiro (2000, 654) allows that special-science taxonomies can "collect and order the domain of a special science in a way that

facilitates its investigation," but that does not mean that such taxonomies contain genuine properties and natural kinds.

What should we make of these closely related challenges to special-
science properties and kinds? It surely depends on the properties and
kinds involved. If it is indeed the case that all samples of jade have little
or nothing in common beyond the fact that they are green, shiny
minerals, and if these properties of jade do not enable us to predict
any other important properties of all and only samples of jade, then we
should conclude that *jade* is not a natural kind, eliminating it in favor of
jadeite and *nephrite*. It would be as though we introduced a category,
golite, including all and only samples of either *gold* or *pyrite* (i.e., FeS_2,
iron sulfide, or "fool's gold"), which share the observable properties of
being shiny, yellow, and metallic. If *jade* is just a disjunction of *jadeite*
and *nephrite*, it lacks projectibility for the same reason that disjunctive
kinds are not projectible in general, as I have already argued (and as is
widely agreed), and there are no further causal properties that all and
only samples of jade have in common. Similarly, if *viscosity* is not a
genuine causal property that features in laws and generalizations, then
we ought to conclude that *viscosity* is not a natural property. And if there
are no further properties that all and only instances of *Newtonian fluid*
have in common (besides constant viscosity relative to applied force),
then *Newtonian fluid* should not be considered a natural kind after all.
But the comparison does a serious disservice to this and other special-
science kinds. We have already seen that the property *viscosity*, through
the diffusion coefficient, features in Fick's first law and plays an explanatory role in a range of different flow and diffusion behaviors in liquids,
gases, and even solids. Similarly, the kind *Newtonian fluid* figures in
important inductive generalizations that are by no means simply deduct-
ive consequences of the property that was used to characterize them in
the first place (namely, that they have a constant viscosity independent
of applied force) and includes both liquids and gases. The flow of
Newtonian fluids can be further described by the Navier–Stokes equation, which applies only to Newtonian fluids, and this equation leads us
to discover further properties of these fluids. Furthermore, as a result of
empirical inquiry, it has been determined that all gases are Newtonian
fluids but not all liquids are. Now, it may yet turn out that *Newtonian
fluid* is only weakly projectible, which would make it a natural kind of a
rather weak sort, since I have already argued that there can be degrees of
naturalness when it comes to kinds (see section 2.7). However, that
would not render it a mere disjunction of two or more basic-science

kinds, with nothing in common to all Newtonian fluids but the single property that was used to characterize it in the first place.

To see this more clearly, consider again Kim's example of the category *jade*, which is a simple disjunction of *jadeite* and *nephrite*. Since the mineral *jadeite* – $NaAlSi_2O_6$ – contains aluminum and the mineral *nephrite* – $Ca_2(Fe, Mg)_5Si_8O_{22}(OH)_2$ – contains calcium, we could conclude: *All samples of jade contain either aluminum or calcium.* But this statement is clearly not a law about jade, since it is a thinly disguised disjunctive statement. A complication arises, however, when we notice by glancing at their chemical formulae again that both jadeite and nephrite (happen to) contain silicon. Now we can say: *All samples of jade contain silicon.* Why is *this* not a natural law, even though it is not obviously disjunctive? Because, at least given what we now know, the presence of silicon has nothing to do with the other properties that jadeite and nephrite share. Unless it could be shown that there were a causal link between the presence of silicon and the macroproperties of jade, it would seem to be a mere accident, not leading to further causal relations that would point to yet other compounds with distinct chemical compositions but similar macroproperties. If it were the case that all compounds containing silicon shared important macroproperties (which is empirically false), then any generalization that we could make about them would apply to all silicon-containing compounds and not just to samples of jade. By contrast, when one identifies a special-science kind like *Newtonian fluid*, one has not just yoked two or more types of compound together artificially. Rather, the substances that we have grouped together turn out to have important properties in common, as witnessed by the fact that there are other things that share those same properties. We are bringing together different phenomena because they conform to certain general causal laws.

Categories like *viscosity* and *Newtonian fluid* earn their keep in science by virtue of their projectibility, and projectibility is an indication that these categories track properties and kinds that enter into real causal relations. Moreover, the projections that they enable us to make are variable in scope. There are generalizations concerning viscosity that apply to all fluids. Then there are generalizations about viscosity common to all gases (e.g., that viscosity *increases* with a rise in temperature) and all liquids (e.g., that viscosity *decreases* with a rise in temperature), but not to all fluids. There are also generalizations about viscosity that are common to all samples of water, but not to all liquids. One could get more specific still and make generalizations about the viscosity of water at a certain temperature. As we narrow our focus, we gain more specific generalizations concerning a more limited range of phenomena. Science is interested in making generalizations about restricted domains but it is also concerned with extending results already obtained to new domains. These two aims are not always in harmony, and a property like *viscosity*, as opposed to its microphysical realizers in distinct domains, enables us to further the latter aim by unifying diverse phenomena and bringing them under the same broad generalizations.

It may be objected that the generalizations that one can make over a wider range of domains are preserved when we narrow our focus, so we do not really give up on anything by restricting our attention to the domain of *liquids* or even more specifically to *water* (as opposed to widening our attention to include fluids more generally, or material substances yet more generally). Generalizations such as Fick's first law, which apply to all *Newtonian fluids*, also (*a fortiori*) apply to all samples of *water*, and to all samples of *water at 20 °C*, and so on. But part of what gives a kind like *Newtonian fluid* scientific importance is that it is applicable across disparate domains. As some philosophers of science have argued, explanatory unification is a cognitive desideratum in science. Even if unification cannot be considered an exhaustive account of scientific explanation, it is at least part of what is aimed at in scientific explanations (Sober 1999, 560). This provides us with a good reason for retaining special-science properties and kinds alongside those of the basic sciences. Kitcher (1995, 172) writes:

[T]he unification of our account of the world is a cognitive desideratum for us, a desideratum that we place ahead of finding the literal truth on the many occasions that we idealize the phenomena. The causal structure of the world, the division of things into kinds, the objective dependencies among phenomena are all generated from our efforts at organization.

Scientists' discovery that they can extend the property of *viscosity* from the domain of liquids to that of gases, and even solids (at least for diffusion phenomena), enables them to extend results already obtained to new domains and to unify what seemed to be diverse phenomena under broader laws and generalizations. This confirms the contention that genuine special-science properties and kinds are a far cry from a disjunctive category like *jade*, which does not appear to be projectible or play an explanatory role, much less one that extends to new domains and phenomena, as *viscosity* was extended from fluids to solids. Moreover, even if special-science properties and kinds are locally reducible to basic-science properties and kinds, this does not render them otiose, since they feature in laws and generalizations that range over different microphysical domains and capture causal patterns that unify diverse phenomena.[3]

3.3 CAUSATION AND SPECIAL-SCIENCE KINDS

The defense of special-science properties and kinds that I offered in the previous section made a distinction between their being multiply realizable and merely disjunctive. The main difference lies in the fact that real properties and natural kinds in the special sciences figure in genuine causal relationships, which enable us to explain and predict natural phenomena. The fact that such kinds can feature in causal relations in their own right is what makes them projectible and what marks them off from nonkinds. In this respect, special-science kinds would appear to be no different from the kinds that derive from the basic sciences. But this conclusion may seem insufficiently attuned to the important differences between the two cases; there are two types of differences that I will focus on in this section and the next. The first is one that has been mentioned already in section 1.7, in a discussion of microphysical fundamentalism. Though skeptics about special-science kinds may acknowledge that some scientific laws invoke them and they appear to enter into scientific explanations, they might say that these laws are not genuine *causal* laws and the explanations are not *causal* explanations, since the real causal work is not being done at the level of the special sciences, but at that of the more basic sciences. Another difference that I will discuss in the next section has to do with the purported fact that laws involving special science properties and kinds tend to be quite different from the laws of the basic sciences.

As we have just seen, Kim and other philosophers have recently argued that what explanatory and predictive value special-science kinds may have must be derived from more basic kinds. Though most philosophers now agree that the laws and inductive generalizations involving special-science kinds are not generally reducible in the classical sense by means of bridge principles to laws of the more basic sciences, there is a widespread consensus that special-science kinds and properties are *supervenient* upon those of the more basic sciences. Since the supervenience relation has been extensively explored in recent philosophical work, I will not dwell

---

[3] The link made here between multiply realizable properties and explanatory unification is related to the account given by Batterman (2000) of multiple realizability. Batterman draws attention to the search for universal behaviors in physics, which ignores the details of microstructure and discovers "features which are stable under perturbation of the micro-details" (136).

on it here, except to say that I will understand the claim in its simplest form – namely, that there can be no difference in macrolevel kinds and properties without a difference in certain microlevel kinds and properties. But Kim and others use this premise to argue that the properties identified by the special sciences are not genuine properties since they lack causal powers. As we shall see, Kim holds that the macrolevel entities studied in the special sciences cannot have causal powers in their own right, since their causal powers are really those of the microlevel entities that constitute them. Where there are no causal powers, there are no properties, and where no properties, no natural kinds. To revert to the example discussed in the previous section, a property like *viscosity* cannot be associated with genuine causal powers, since the causal powers would appear to be possessed by the microparticles that constitute each specific viscous substance or medium. Hence, special-science properties are not real properties and special-science kinds are not genuine kinds. The *predicates* that we use to identify properties and kinds at the macrolevel may well have explanatory efficacy and predictive value, but they do not pick out real properties and natural kinds.

To respond to this general worry about causation, we need to take another look at the details of the case that was discussed in the previous section. Let us suppose that we attempt to achieve a functional reduction of *viscosity* in liquids to a lower-level scientific property. The basic-science property involved in this case might be expected to pertain to the microparticles of liquids, generally molecules. Clearly, *viscosity* cannot be a property of individual liquid molecules since we have already seen that it is primarily dependent on the bonds *between* the molecules and also that the phenomenon of fluid flow can only pertain to large *collections* of molecules. As a first stab and for the sake of simplicity, we might say that the basic-science property that corresponds to *viscosity* in a liquid has turned out to be a function of the strength of the chemical bonds between the molecules in a liquid: The stronger the bonds the higher is the viscosity, and the weaker the bonds the lower the viscosity. Molecules of a liquid, in sufficient numbers, and taking into account the bonds between them, would seem to determine the viscosity of a sample of the liquid as a whole. However, talk of the "strength" of the bonds is a huge oversimplification, since chemical bonds come in various different types, such as covalent, ionic, and dipole, and "strength" is dependent on different properties in each case. But viscosity also depends on temperature, so it is not enough to take into account the chemical bonds between the molecules – one must also factor in the kinetic energy of the molecules involved. Moreover, the dependence on temperature is not a mere background condition that can be factored out, since it makes different contributions in different cases. Some liquids show relatively constant viscosity with change in temperature while others show more variability in viscosity due to temperature, and the relationship is not always linear. In addition to the strength and nature of the chemical bonds between molecules and the kinetic energy of the molecules, we have already seen that in non-Newtonian liquids, viscosity also depends on the applied force. Hence, the *viscosity* of a fluid does not merely correspond to a set of intrinsic properties of the molecules of the fluid itself but is largely determined by the relations of those molecules to each other as well as to external conditions, such as an applied force. Initially, it may have been tempting to think of a property like *viscosity* as pertaining intrinsically to a type of substance, such as water, oil, or air, and as inhering in the molecules of that substance as an intrinsic property or set of properties of those types of molecules themselves. However, not only does viscosity depend on the relations between those molecules, the phase of matter they are in, and the kinetic energy that those molecules have, but it also depends in some substances on the external force that is applied to them and the rate at which that force is being applied. From the perspective of the microlevel, *viscosity* is a largely extrinsic affair, pertaining not to molecules themselves but to their configurations, relations, energy states, and the nature of the forces that are acting on them. This means that even talk of a *local* reduction of *viscosity* to the microproperty of a kind of chemical compound, is unrealistic, depending as it does on a variety of extrinsic macrolevel conditions.

The *viscosity* of a fluid is the result of a complex array of intrinsic and extrinsic factors that come together in certain ways to issue in, or realize, a certain macroscopic phenomenon. Two substances can have the same viscosity as a result of very different configurations at the microlevel; moreover, samples of the same chemical substance may have different viscosities due to different circumstances – say, a difference in temperature. For example, two samples of engine oil may have different chemical formulae yet have identical viscosities due to different values of temperature; meanwhile, two samples of oil may be identical in chemical formula yet have different viscosities because they are at different temperatures. Since *viscosity* is an effect of multiple causes, some distinct causal process may be sensitive only to the effects of these causes, treating equivalent effects in the same way and issuing in identical outcomes, regardless of the underlying particles, configurations, and so on. Consider a mechanical engine running on low-viscosity engine oil, rendering it more fuel-efficient. Suppose we have two cars both of which run on low-viscosity engine oil; in one, the oil has low viscosity because of its chemical composition and the temperature, while in another the substance has the same low viscosity because of a different chemical composition and a different temperature. They end up with the same value for viscosity for different reasons and in virtue of different chemical compositions and different temperatures. If we pose the question, "What caused a 2% reduction in fuel consumption in the two cases?," a reasonable answer would be the following: "The same low value for the viscosity of the engine oil caused the same level of fuel reduction in the two engines." For the purposes of this explanation, the underlying account about chemical composition and temperature may not be relevant.

It might be objected here that explanation is not causation. But these are evidently causal explanations that purport to appeal to real causal relations, and if Kim's worries were warranted, then the causal claims in such explanations would surely be spurious. At least prima facie, we have given an explanation that cites a cause. Yet, despite the fact that some scientific explanations do not require us to appeal to the microlevel to understand similar or different values for viscosity, there remains a metaphysical conviction among some philosophers that whatever explains these phenomena, a complete causal account of them must appeal ultimately to the microphysical goings-on, and, more importantly, that causation at the microlevel necessarily "excludes" causation at all levels above the microlevel. The "causal exclusion" argument has been discussed extensively in recent philosophical work, and what I have to say about it will be brief. Kim has been the main exponent of the argument, and, as we have already seen, he characterizes the relation between microphysical and macrophysical levels in terms of *realization*: Microlevel properties realize macrolevel properties. Though he usually explicates the relation of supervenience for mental properties and physical properties, he makes it clear that he intends it to apply more generally to special-science properties and the basic-science properties that realize them. Here is a relatively informal way of presenting the "causal exclusion argument" for mental properties, M and M^*, and their physical realizers, P and P^* (cf. Kim 1998, 37–47, 2003, 155–159). Take any case in which an instance of mental property M purportedly causes an instance of another mental property M^*. M has a physical supervenience base P, and M^* has a physical supervenience base P^*, which means that on this occasion an instance of P also causes an instance of P^*. Hence, M causes M^* only by virtue of the fact that it causes P^*, the supervenience base of M^*. But now it seems that both P causes P^* and M causes P^*. However, there is a metaphysical principle that says that the physical realm is *causally closed* in the sense that every caused[4] physical event or property instance must have a *physical* cause. Since no event or property instance can have more than one sufficient cause (except in cases of causal overdetermination), it follows from causal closure that it must be P that really caused P^*, not M. Generalizing from this case, it seems that no mental property instance can ever act as a cause, so there is no genuine mental causation. And generalizing even further, no higher-level property that supervenes on a lower-level property can ever be causally efficacious, so there can be no higher-level causes and properties. Clearly, the argument rests crucially on the premise that no property instance can have more than one sufficient cause; in particular, where a macrolevel property is realized by a microlevel property, there cannot be both macro- and microcausation. Kim (2003, 157) articulates the "causal exclusion principle" as follows:

[Exclusion] No single event can have more than one sufficient cause occurring at any given time – unless it is a genuine case of causal overdetermination.

If some property instance P is the realizer of some property instance M, then the causal powers instantiated by M are none other than the causal powers instantiated by P.

In section 1.7, in arguing against the microphysical fundamentalism posited by some essentialists, I put forward two considerations for doubting this argument. One consideration is that the argument presupposes that there is some "bottom" level where causation must actually take place.[5] Yet, if there is no such bottom level and levels descend infinitely (which is a distinct empirical possibility), then the argument seems to imply that there is no causation anywhere. For all we know (and may ever know), the universe we live in is so constituted, but if it is, I argued that we should not deny the existence of all causal relations. And if it would be wrong to conclude that there is no causation in the "bottomless pit" case, then we cannot consistently say in the case that there *is* a most fundamental level that genuine causation takes place only at the bottom-most level.

---

[4] The qualification is necessary to allow for uncaused quantum events, like the radioactive decay of an atomic nucleus.

[5] Kim (2003, 173) writes: "As I understand it, the so-called Standard Model [of elementary particle physics] is currently taken to represent the bottom level. Assume that this level is causally closed; the supervenience argument, if it works, shows that mental causal relations give way to causal relations at this microlevel."

A second consideration against this argument is that, assuming that there is a fundamental level, consistent application of the argument across the board would leave us with only quarks and leptons; no atoms, molecules, cells, neurons, and so on. This means that there would be no causation in any science but elementary particle physics, an outcome which is problematic because the special sciences are up to their necks in causal claims, as we just saw in the case of the causal explanations citing *viscosity*, which philosophers cannot lightly dismiss.[6]

These considerations against Kim's causal exclusion argument may not seem decisive and the second one, in this context, may appear to beg the question. But they are pitted against an argument that appeals to at least one principle, the causal exclusion principle, which is usually stated as a bare axiom in the argument and does not receive adequate justification. Although it is taken to be obviously true by some philosophers, and is indeed considered "analytic" by Kim, it is squarely denied by others.[7] The dissenters have not just been content with flat rejection of this principle; some have argued that on at least some analyses of causation the causal exclusion principle comes out as false. It is possible either for there to be causation at both micro- and macrolevels where properties supervene on one another, or for causation to involve macrolevel events *to the exclusion of the microlevel events* upon which they supervene. This argument has been made with reference to several different analyses of the concept of causation, notably the interventionist account of causation proposed by Woodward (2003). In this vein, List and Menzies (2007) use an insight from Yablo (1992) that causes must be proportional to their effects, neither too specific nor too general. If a pigeon pecks at a scarlet target, the cause of its pecking may be that the target was red, not that it was scarlet (even though it was scarlet), since it would have pecked just as long as it was red. In a similar fashion, to say that P caused M^* may be too specific if some other neural configuration P' would have caused it too. They acknowledge that this depends on a counterfactual or interventionist understanding of causation (rather than a production or generation one, which Kim favors).[8]

---

[6] This is not to suggest that Kim dismisses it *lightly*, but, on balance, he regards his metaphysical argument as forcing us to accept this conclusion. He writes (2003, 172), somewhat wistfully: "It would be nice if we could embrace causation at many levels, including the psychological, the biological, and so on, and also cross-level causation, both downward and upward, all of them coexisting in harmony."

[7] Kim (2003, 163) states: "Although the causal exclusion principle has been widely accepted and I believe it is virtually an analytic truth with not much content, some find it problematic."

[8] For similar arguments that utilize the interventionist analysis of causation, see List and Menzies (2009), Marras (2007), and Raatikainen (2010). See Elder (2004) for an argument to this effect for

But since discussion of these points about different conceptions of causality would take us too far afield, I will not pursue them here, except to note that this is one juncture in our investigation of natural kinds in which we seem to have arrived at an impasse caused by a conflict between commitment to a certain metaphysical principle (on the one hand) and a naturalist acceptance of explanatory practices in the sciences (on the other). In Chapter 1, I argued that the methodology that I would pursue in this inquiry would aim to reach reflective equilibrium between our metaphysical commitments and the findings of scientific inquiry. How should we adjudicate the current standoff? The deck seems stacked against the causal exclusion principle. For not only does that principle not receive independent justification beyond its immediate intuitive appeal to some philosophers, but it is also countered by metaphysical arguments to the effect that some analyses of causation either are not committed to it or explicitly undermine it. Given that explanatory practices in the special sciences, at least when taken literally, appear to flout this principle on a regular basis, it seems wise not to accept the principle and with it the conclusion that there is no macrolevel causation, and hence no macrolevel properties and kinds.

3.4 NATURAL LAWS AND SPECIAL-SCIENCE KINDS

In the account of natural kinds that I have been defending, I began by according prominence to the notion of projectibility. Then I went on to argue that projectibility is a diagnostic feature of natural kinds, reflecting the fact that the properties associated with natural kinds are involved in causal relations with one another. What enables us to project from one property associated with a natural kind to another, or from one natural kind to another, is the fact that they enter into causal relations with one another. Our initial characterization, which had it that each natural kind K is associated with a set of properties $\{P_1, P_2, \ldots, P_n\}$, where the properties are "important" from the point of view of a branch of science and where K and the P's are projectible, did not sufficiently incorporate the *causal* dimension of natural kinds. As we homed in more closely on the causal dimension of natural kinds, it became clear that some of the properties associated with natural kinds enter into causal relations with other

the INUS analysis of causation, according to which a cause is an insufficient but necessary part of an unnecessary but sufficient condition. Meanwhile, Horgan (2001, 102) holds that "concepts of causation and causal explanation are contextually parametrized notions, with an implicit contextual parameter keyed to a specific descriptive/ontological level." He calls this position "causal compatibilism."

properties that are so associated. Rather than simply consider a natural kind as being associated with a set of properties $\{P_1, P_2, \ldots, P_n\}$, it is more accurate to consider a natural kind as being typically associated with a subset of those properties, which have causal priority over the others and either singly or jointly give rise to another subset of properties, which may in turn give rise to others. In both cases, the subsets of properties may be loose clusters rather than being singly necessary and jointly sufficient for membership in the kind. Some of these properties will be quantitative and determinate (rather than determinable) and will have a certain definite value or range over a set of values. Moreover, the Ps associated with natural kinds are often complex functions of simpler properties. For example, we have seen that the kind *Newtonian fluid* is related to the property *viscosity* in the following way: A Newtonian fluid is one whose viscosity remains constant with a change in the applied force. This property is a complex function of two simpler properties, *viscosity* and *force*. Since the causal relationships among the properties associated with natural kinds are capable of being captured in laws or generalizations, these kinds are eminently projectible. This, in a nutshell, is the "simple causal theory" of natural kinds that I alluded to in Chapter 2. Presumably, there is a close relation between the causal dimension of natural kinds and their featuring in scientific laws, for many scientific laws are overtly causal or reflect causal relationships. Since some philosophers have questioned the status, and indeed the very existence, of laws in the special sciences, I will now take a closer look at this issue. Are laws in the special sciences fundamentally different from those in the basic sciences? Do the alleged limitations or shortcomings of special-science generalizations imply that the categories that they contain do not correspond to natural kinds and real properties?

In the previous section, we encountered a number of general statements from the special science of fluid mechanics, all of which involve the property *viscosity* or the kind *Newtonian fluid* (or both):

(1) Fick's first law: $J = -D \, (\delta C/\delta x)$, where $D = k_B T/6\pi\eta$
(2) All Newtonian fluids have a viscosity that is independent of applied force.
(3) All gases are Newtonian fluids (but only some liquids are Newtonian fluids).
(4) In gases, viscosity increases with temperature.
(5) In liquids, viscosity decreases with temperature.
(6) Water is a Newtonian fluid.

(7) The viscosity of water at 20 °C is 1.002 cP (centipoise, where 1 poise is 0.1 pascal second, or 0.1 Pa s.).

Some of these generalizations may not be true laws of nature, but is there some principled reason for thinking that *none* of them are? These generalizations are not all on a par. For one thing, only some have the explicit form of general propositions or universally quantified statements, though they can all be put in this form without much effort (e.g., *All samples of water are samples of Newtonian fluids*). More importantly, most of them are at least partly grounded in causal facts of one sort or another, even those that do not refer overtly to a causal process (e.g., (6) can be paraphrased to describe what happens to the viscosity of water when force is applied to it). Some of them, like (1), are quantitative laws that relate properties to one another in the form of a mathematical equation, while others, like (3), are qualitative, relating natural kinds to one another in a systematic way. Further, (2) is arguably not a law since it serves to introduce a kind, *Newtonian fluid*, and associate it with a complex property, rather than inform us further about a kind that has already been introduced. Still, I would argue that (2) is not analytic in the sense that it is devoid of empirical content for familiar reasons – namely, that even such a statement is not immune to revision in the context of empirical inquiry. As we might expect when it comes to laws and generalizations in the special sciences, all of them appear restricted in their domain of application. In some of these, the restriction is obvious since the antecedent explicitly specifies its range; for instance, (4) states that it is a generalization about gases. In others, this needs to be further specified; for example, (1) pertains primarily to fluid flow or diffusion. Moreover, some of these statements may seem too narrow at first sight to qualify as laws, such as (7), which states the value for the viscosity of water at 20 °C. Nonetheless, it tells us about *all* samples of water at this temperature and enables us to predict how other samples of water will behave under certain specified conditions. It is what is sometimes called a "phenomenological law," since it is established on the basis of observation or experiment rather than derived from theory. I will now consider and assess a number of reasons for thinking that the generalizations of the special sciences cannot be statements of natural law, or that they are crucially different from the laws of the basic sciences.

Natural laws are not just true (or nearly true) empirical generalizations – they are also supposed to have modal force. Many philosophers have regarded it as a central if not a defining feature of natural laws that they are (nomologically) necessary rather than merely contingent, or that they can be used to support counterfactual statements. Though I am skeptical of claims of necessity and contingency simpliciter and think that these judgments are generally context-sensitive and made relative to background assumptions, it would appear that special-science generalizations are no worse off than basic-science laws in this respect. Just as we can say that the law, *All protons have a charge of 1.6×10^{-19} C*, licenses the inference (at least in some contexts) that, *If that had been a proton (rather than a neutron), then it would have had a charge of 1.6×10^{-19} C, and would have formed a stable atom with an electron*, we can also say that the generalization, *All water at 20 °C has viscosity 1.002 cP*, licenses such inferences as, *If this had been a sample of water (rather than ammonia), then it would have had viscosity 1.002 cP, and it would have been more resistant to flow*, or *If this sample of water had been at 20 °C (rather than at 10 °C), then it would have had viscosity 1.002 cP, and would have been less resistant to flow*. Thus, special-science generalizations are not accidental statements, and there is no obvious difference between at least some of these statements and those of more basic sciences when it comes to their modal force.

What about other putative differences? Natural laws come in many shapes and sizes and can differ along various dimensions; for instance, range of application, strictness, or exceptionlessness; whether they capture a quantitative relationship between properties or the values that those properties can take (e.g., proportionality between values of P_1 and values of P_2) or merely the coincidence of properties and natural kinds (e.g., whenever there is P_1 and P_2, there is P_3); and whether they are strict or probabilistic, among other dimensions (cf. Mitchell 2000). On some of these measures, especially range of application and strictness, the laws of the basic sciences are thought to fare better than the generalizations of the special sciences. This has led some philosophers of science to restrict laws to the basic sciences, but a more common view is to regard them as lying on a continuum (or several continua) when it comes to these dimensions, from laws that are applicable in all systems in the universe across the entirety of space-time to those that are applicable only in certain regions in the universe for certain restricted systems, from exceptionless laws to exception-ridden ones, and so on.[9]

---

[9] For such a nondichotomous view of natural laws, see Mitchell (2000), on which this paragraph draws, as well as Hitchcock and Woodward (2003). Mitchell advocates what she terms a "pragmatic account" of natural laws, which emphasizes the role that they play in the sciences and the function that they serve.

Does the restricted range of application of special-science laws dis-qualify them from being regarded as true laws of nature? The assump-tion embedded in this question can itself be queried. Though I have been speaking of the explicit or implicit restrictions found in these statements, there is a sense in which at least some of them are not actually restricted. The generalization, *All gases are Newtonian fluids*, can be' regarded as universal. It applies only to gases, true enough, but it does so everywhere in the universe. The same goes for, *Water is a Newtonian fluid*, which applies wherever water is found. If this sounds like a verbal trick, consider what is uncontroversially thought to be an unrestricted or universal law statement, *All protons have positive charge or 1.6×10^{-19} C*. It could be argued that this statement is also restricted in scope since it applies to all and only protons. In a region of the universe where there are no protons, it would have no application. But that would be perverse since it is clear that it would apply to protons wherever they might be. However, the same could be said about the statements concerning gas and water, respectively. Is the difference that protons are more pervasive throughout the universe than gas, which is more pervasive than water, and so on? But then saying that the first two statements are restricted in scope while the third is not would appear to be a matter of arbitrarily drawing a line somewhere based on the relative prevalence of protons, gases, and water.

The above points about the comparative range of these laws or generalizations may seem to be a consequence of the fact that I have chosen a *physical* special science as my case study rather than a special science drawn, say, from the biological or social realms. But similar points apply when comparing generalizations drawn from the nonphy-sical special sciences with those of fundamental physics. Although they are realists about the properties and kinds of the special sciences, Ladyman and Ross (2007, 195) nevertheless differentiate between the generalizations of the special sciences and those of fundamental physics, as follows:

[A] science is special if it aims at generalizations such that measurements taken only from restricted areas of the universe, and/or at restricted scales are potential sources of confirmation and/or falsification of those generalizations. For example, only measurements taken in parts of the universe where agents compete for scarce resources are relevant to the generalizations of economics, and no measurements taken in the center of the Earth are relevant to psychology.

By contrast, they say that "the locator 'top quark' directs us to take measurements everywhere in the universe" (2007, 122).[10] But as long as there are some regions of the universe in which there are no top quarks, it is not the case that measurements taken anywhere in the universe are relevant to generalizations about top quarks; indeed, top quarks are very rare in the universe. It is not even the case that a top quark *could* in principle be at any location in the universe (whereas economic markets and psychological agents could not be), since a top quark could not be bound to a top antiquark, on pain of annihilation. Moreover, only measurements taken "at restricted scales" are relevant to generalizations about quarks; otherwise, elementary particle physics could rely on macrolevel measurements rather than having to resort to linear accelerators. Again, we seem forced to fall back on a vague notion like the relative abundance of quarks (on the one hand) and economic markets and psychological agents (on the other), rather than the relative scope or range of the generalizations that apply to their respective realms. Protons, gases, and water are more plentiful in the universe than biological organisms, psychological agents, and economic markets. But that fact does not give us grounds for ruling that generalizations about the latter are restricted while those about the former are unrestricted in scope.

It is true that among the generalizations I have listed from fluid mechanics, some have a wider range of application than others; for instance (5), which describes the viscosity of all liquids with respect to temperature, is less restricted than (7), which deals specifically with the viscosity of water. But direct comparisons seem possible only when we are discussing statements drawn from the same scientific domain. When it comes to distinct scientific domains, there does not seem to be a straightforward way of comparing their range of application and ruling that some are less restricted than others.[11] Hence, a principled distinction between special-science generalizations and laws of nature on the basis of the allegedly restricted character of the former, or their limited range of application, does not appear to be forthcoming.

Another feature that has often been attributed to special-science generalizations, which should not be confused with their alleged restricted range or scope, is that they are commonly said to admit of exceptions rather than be exceptionless. This feature can be captured by saying that they contain

[10] In Ladyman and Ross' terminology, a "locator" is "an act of 'tagging' against an established address system," and an "address system" is roughly a representational system (2007, 121).

[11] In section 3.5, I will say more about the notion of "domain" that is being used here.

explicit or implicit *ceteris paribus* clauses (e.g., "other things being equal").
Now consider (5), which says that the viscosity of a liquid decreases with
an increase in temperature. This statement has to be qualified in all sorts of
ways to allow for exceptions. For example, if the liquid is a *non-Newtonian
fluid*, we also need to say that this statement holds only if the applied force
is held constant, since in such fluids an applied force changes the value of
viscosity. Moreover, there are doubtless additional qualifications that
would need to be added to ensure that other factors do not influence the
viscosity, and we may not be able to anticipate the myriad ways in which
this general statement may have to be qualified for specific liquids. Our
inability to specify all the spoilers for this generalization leads to the
following familiar dilemma. On the one hand, we may attempt to handle
the various unanticipated qualifications by including a catch-all rider or
proviso, such as "other things being equal" or "provided there are no other
sources of interference". But this amounts to saying that the viscosity of
liquids increases with temperature unless something happens to prevent it,
in which case it does not. This trivializes the initial statement, which now
effectively says either that the viscosity of liquids increases with temperature or it does not. On the other hand, if we leave this qualification out,
and simply state that, for all liquids viscosity increases with an increase in
temperature, then the statement is literally false (since in some cases it
does not). Even if we add the known qualifications, as long as there are
others that we have not anticipated (and a little speculation suggests that
there will be others, at least in some cases), then the statement remains
false. This dilemma casts a shadow on special-science properties and
kinds, since they are in danger of appearing not to track robust causal
relations. If nothing can be said that is both nontrivial and true about the
way in which viscosity varies with temperature in liquids, then in what
sense is *viscosity* a real property that reveals something about causal
relations and natural phenomena?

Philosophers of science have confronted this dilemma in a number of
ways. One promising strategy for dealing with the problem begins by
observing that many generalizations in the special sciences and basic
sciences alike summarize causal relations between different properties or
property instances (e.g., *viscosity* and *temperature*) contingent upon certain
background conditions. Moreover, the difference between the special-
science and basic-science generalizations lies primarily in the conditions
upon which the causal relations are contingent; the conditions upon which
causal relations in the basic sciences are contingent are stable relative to
those in the special sciences. For example, Mitchell (2000) states that the

difference between Galileo's law of free fall (which is contingent on the mass of the Earth) and Mendel's law of segregation (which is contingent on the existence of sexually reproducing biological organisms) can be understood in terms of the degree of stability of the conditions under which they hold. But since it is a matter of degree, the difference between them is not absolute. After discussing a range of other examples, Mitchell (2000, 254) writes:

At one end of the continuum are those regularities whose conditions are stable over all time and space. At the other end are the so-called accidental generalizations. And in the vast middle is where most scientific generalizations are found.

Galileo's law may not be the best example from our perspective, since, according to the criteria I have adopted, it is more plausibly regarded as a special-science law. But as other philosophers have pointed out, even the most fundamental basic-science laws may not hold completely without exception across all space and time. The second law of thermodynamics can be violated in nonequilibrium states of brief temporal duration. The law of mass–energy conservation may not hold across a spatial singularity (e.g., a black hole) or a temporal singularity (e.g., the big bang).[12] Obviously, these laws are far more stable (in Mitchell's sense) than the laws of the special sciences, but stability is relative and it does not seem to mark a fundamental rift between special-science and basic-science generalizations.[13]

There may be other differences between the exceptions to special-science generalizations and the exceptions to basic-science generalizations. In the case of the special sciences, the exceptions cannot always be anticipated and may be too numerous to list. When we can anticipate them and understand the reasons for them, they are often due to interactions that can best be explained at another level of description. In the case of the general statement that viscosity decreases with an increase in temperature in liquids, one exception involves the element sulfur, whose viscosity increases at a certain temperature between its melting and boiling points because polymerization occurs and there is a change of allotrope.

---

[12] Compare Ross (2000, 155); see also Karbasizadeh (2008, 19–20).

[13] Cartwright (1999) argues on somewhat different grounds that all laws, including the most fundamental laws of physics, are *ceteris paribus* laws and that none are unconditional and unrestricted in scope. In her view, laws only ever hold in the context of a "nomological machine," which is "a fixed (enough) arrangement of components, or factors, with stable (enough) capacities that in the right sort of stable (enough) environment will, with repeated operation" give rise to regular behavior (1999, 50).

This exception can be explained by referring to microlevel reality. Crudely put, the nature of the chemical bonds change in sulfur at a certain temperature, making the bonds stronger and rendering the liquid more viscous with an increase in temperature (but at yet higher temperatures its viscosity decreases again). If explaining the exceptions to macrolevel generalizations involves invoking processes at another level of description, it may seem reasonable to suppose that laws that apply at the most fundamental microlevel are strict and exceptionless, and this is what sets apart basic-science generalizations from those in the special sciences. But now consider the law that all elements in group 18 in the periodic table are noble gases (e.g., helium, neon, argon, and so on) due to the fact that they have no valence electrons and their outermost electron shell is full. The only known exception is the heaviest element produced in the laboratory to date, atomic number 118, which is also classified in group 18. This exception to the law is thought to arise as a result of relativistic effects. Given the size and complexity of this atomic nucleus relative to the others, interactive effects arise that are not present in simpler nuclides classified as noble gases. Hence, even at what seems to be the most fundamental level, there are interferences that lead to exceptions to what are thought to be strict laws. But perhaps it may be said that the reason for there being exceptions in this case is that this is not the most fundamental level and that if we really descend to the level of quarks and leptons (or whatever the bottom level turns out to be, if there is one), then we would find truly exceptionless laws. If we grant that there is a most fundamental level, then it may be that there will be absolutely strict laws that apply at that level. Perhaps the law of conservation of charge is a natural law of this kind. But we have already seen that even such a principle as mass–energy conservation does not apply universally and without exception, since it can be violated due to interference from quantum effects, as in the vicinity of a black hole. If some fundamental-level generalizations are exceptionless while others admit of exceptions, then we cannot use this to mark a fundamental difference between the basic and special sciences. Generalizations in the special sciences are not exceptionless because macrolevel systems studied in the special sciences tend to be larger and more complex, and have more parts than those studied in the basic sciences. This means that there will be more opportunity for interference with these macrocausal processes by other causal processes, both macro- and micro-, as chemical polymerization interferes with fluid flow. Hence, their generalizations will tend to have more exceptions than those in the basic sciences. But the difference is apparently a matter of degree. This feature of special-science laws or generalizations accords well with the observation that the causal connections among many of the properties associated with special-science kinds are not always strict. Rather than strict causal relations among the properties associated with a natural kind in the special sciences, we often observe loose correlations among them, which is to say that the natural kinds involved often enter into generalizations that admit of exceptions rather than strict or exceptionless laws.

When it comes to the other dimensions along which laws may differ, there do not appear to be stark differences between generalizations in the special sciences and the basic sciences. It is clearly not the case, based on some of the examples we have encountered, that special-science generalizations are qualitative rather than quantitative. This applies not just in disciplines such as fluid mechanics but also in such special sciences as population biology and economics, where generalizations are frequently framed in quantitative terms and describe precise mathematical relationships between properties. Furthermore, it is false that special-science generalizations are probabilistic or statistical while basic-science laws are nonprobabilistic or deterministic. For one thing, quantum mechanical laws such as the Schrödinger equation are fundamentally probabilistic. But it may be objected that this kind of probability is metaphysical rather than epistemic: It is a feature of the nature of reality rather than a reflection of our ignorance (at least according to the dominant interpretation of quantum theory). This marks a genuine difference between quantum mechanical probabilistic laws of the basic-sciences and those of the special sciences. (But it is not clear that we can say the same about the second law of thermodynamics, which may or may not be metaphysically probabilistic, depending on how fundamental it turns out to be.) However, the claim that special-science generalizations are epistemically probabilistic while basic-science laws are (at worst) metaphysically probabilistic seems to take us back to a consideration of exceptions. If the probabilistic character of some special-science generalizations is a reflection of our ignorance, then it is presumably so because we are not aware of all the possible ways in which other processes might interfere with the process that the general statement describes. Even if we cannot enumerate all these sources of interference, we may be able to quantify them. Indeed, this is one way in which we can tame exceptions to a scientific law or generalization,[14] since instead of attaching a catch-all *ceteris paribus* clause, the general statement itself will be said to apply only in a certain

---

[14] For other ways in which *ceteris paribus* laws can be rescued from vacuity, see Kincaid (1990, 72).

proportion of cases or with a certain frequency. If it is indeed the case that many, or indeed all, special-science generalizations are epistemically probabilistic, this is tantamount to the claim that many (or all) are prone to exceptions. But we have already dealt with that feature and have seen that the same can arguably be said for at least some of the most fundamental laws of physics. Now that we have explored the main dimensions along which scientific laws or generalizations have been claimed to vary, the conclusion that has emerged is that there is no hard-and-fast distinction between special-science and basic-science generalizations on any of these dimensions. At best, there is a difference between the generalizations of the special sciences and *some* of the generalizations that occur in the basic sciences, when it comes to the exceptions that they admit. Hence, special-science kinds and properties cannot be starkly differentiated from basic-science kinds and properties on account of the nature of the respective generalizations that we can make about them.

3.5 REAL CAUSAL PATTERNS

The existence of natural kinds in the special sciences depends on there being genuine causal relations in their domains, which relations can at least sometimes be captured by empirical generalizations that are similar in many respects to natural laws, albeit ones hedged with exceptions or *ceteris paribus* clauses. The chemical kind *water* is related in this way to the fluid mechanical kind *Newtonian fluid*. The truth of the generalization, *All samples of water are Newtonian fluids*, depends on the fact that the viscosity of water does not change with an applied force (at a given temperature). *Water* is a natural kind,[15] one of whose associated properties is a certain viscosity, and another of its properties is that it has constant viscosity with applied force. *Newtonian fluid* is a natural kind because it is projectible; members of this kind all maintain a constant viscosity if the force applied to them changes, and they participate in causal processes that can be described by the same laws of fluid flow.

---

[15] See Weisberg (2006) for an argument that there are several chemical kinds corresponding to *water*, depending on the theoretical purpose or the context of the theoretical inquiry. Any of these several chemical kinds will do for my purposes here, since all of them are nomologically related to the kind *Newtonian fluid*, though the viscosity of each of them will be slightly different. Alternatively, it may be possible to construe water as a fuzzy kind with graded structure, like the kinds of isomeric chemical compounds discussed in section 2.4. Moreover, note that this kind (or kinds) is best understood as a macrolevel kind and cannot be identified with a single microlevel chemical structure such as H_2O.

The projectibility of these kinds and properties is a good sign that they are natural kinds, and the regular and repeatable causal relations that they enter into are what make them projectible. Some of these causal relations can be captured by precise mathematical equations, such as Fick's first law, while others are expressible in the form of nonquantitative generalizations. Although many (if not all) such generalizations must be qualified with *ceteris paribus* conditions, that does not prevent them from being natural laws. Recall that I argued that special-science properties and kinds enable us to unify a diverse range of phenomena and to make generalizations across disparate domains. These generalizations cannot be captured simply by disjoining the generalizations that can be made about each of these domains, since there is no limit in principle to the range of phenomena that fall within their scope. Fick's first law of diffusion is a case in point: It was applied originally to fluids (liquids and gases) but was later shown to apply when a liquid diffuses through some solids in case those solids are porous. Any attempt to capture the overarching law using a disjunction of microlevel laws (even if we had a complete local reduction in each type of system, or type of substance) would fail to capture the full generality of Fick's first law.

This drive for generality or unification in science plays out in a surprising way when it comes to the property of *viscosity*. We have seen that the property of *viscosity* pertains to liquids, gases, and some porous solids, and that Fick's first law of diffusion applies in all of these cases. But it turns out that one can define *viscosity* even more generally. The biologist W. D. Hamilton (1964, 10) proposed characterizing a "viscous population" of biological organisms as one marked by "a general inability or disinclination of the organisms to move far from their places of birth." According to this now widely accepted proposal, a population of organisms that are reluctant to stray from their birthplace or other initial location (for one reason or another) is more viscous than a population in which organisms are more mobile. Other things being equal, genes will flow more slowly through a highly viscous population than one that is not, and it turns out that some of the same patterns that characterize fluid flow can also be used to characterize genes flowing through a population of organisms over generations. The property of *viscosity* and Fick's first law of diffusion have been used to describe changes in allele frequency in a population, especially in social insects.[16] If we view genetic transmission in terms of fluid flow and

---

[16] To get a flavor of such work, consider the following statement regarding two species of red ants: "If polygynous colonies recruit their own daughters as new reproductives and polydomous (multinest)

diffusion, we will be able to transfer some of the results already established for fluids to this case (while we also give up on others, which apply only to material fluids). The fact that the concept of *viscosity* crops up in hitherto unanticipated domains and figures in causal relations in these various domains, demonstrates the way in which some of the same types of causal processes recur across diverse contexts. This vindicates science's interest in unifying seemingly diverse phenomena through categories that enable us to frame broad generalizations that transcend what are intuitively regarded as single domains. There is no reason to suppose that there is a limit on the situations in which the property of *viscosity* is instantiated, and in which it enters into some of the same causal relations and satisfies some of the same empirical generalizations. This distinguishes a genuine scientific property from a mere disjunctive category.

But now the following worry might emerge. If we broaden the category *viscosity* to such an extent that it can be correctly applied to a population of biological organisms, then we may have overstretched. As long as the category was restricted to fluids (and possibly solids), it seemed as though we were tracking patterns of causal relations that were fundamentally the same in the various domains – hence, real properties and natural kinds. But when the same category is applied to phenomena in which there is no material substance flowing through a physical channel or diffusing through another material substance in a literal way, the extension may seem to be based on a loose analogy. To press the point further, could the category of *viscosity* or Fick's first law be properly applied, say, to the flow of vehicular traffic on a highway, or to cultural innovations diffusing through a human population across generations, or to the interstellar diffusion of alien galactic civilizations?[17] Would these be legitimate extensions of the same real properties and natural kinds as are found in the other domains we have discussed, or are they distant analogues of the original phenomena? In other words, are we still discerning real causal patterns in the universe or

---

societies arise through budding, allele frequencies may cluster spatially, and the population become viscous" (Seppä and Pamilo 1995, 201).

[17] This is not pure speculation on my part. The property of *viscosity* has been deployed by transportation engineers to model the flow of vehicles; Zhang (2003, 30) writes: "[I]n traffic flow it is the driver's tendency to resist sharp changes in speed that may lead to viscous terms in traffic speed dynamics." Meanwhile, Fick's law has been invoked by researchers investigating the possibility of galactic colonization by alien civilizations. Based on their diffusion models of expanding galactic populations and the lack of contact to date with extraterrestrials, Newman and Sagan (1981, 293) conclude somewhat reassuringly that "except possibly in the early history of the Galaxy, there are no very old galactic civilizations with a consistent policy of conquest of inhabited worlds."

are these merely loose analogies across fundamentally different domains? If we are too lax in judging that the natural kind is manifested in new domains, this might seem to undermine the reality of the category in question and cast doubt on the whole notion of real properties and natural kinds in the special sciences.

This challenge raises an important question concerning the limits of explanatory unification in science. The response to it should surely depend on the extent to which the same quantitative and qualitative relationships can be tracked in the separate domains. To the extent that *viscosity* in population genetics enters into the same causal relations and generalizations as its fluid counterparts, it is legitimate to say that we are talking about the same special-science property. But if the causal relationships among phenomena are found to be too loose to sustain explanations and predictions in the new domain, then we will have to conclude that extending them to some new set of phenomena is merely a loose analogy. If the disanalogies pile up and outweigh any explanatory benefits to be gained from extending the concept, then we would be justified in saying that the concept does not really apply, and hence that the property or kind is not actually manifested in these novel arenas. For example, it is unlikely that we will be able to identify the equivalent of a *Newtonian fluid* in the domain of population genetics because there does not seem to be an obvious analogue to the notion of an applied force within a population of organisms passing their genes to successive generations. Since we are talking about the extent and strength of certain causal relations, there is unlikely to be a simple answer to the question of whether a real property or natural kind is manifested within a given range of phenomena. When the causal patterns are stable enough and similar enough to warrant it, a scientific subdiscipline or independent special science often arises to study the similarities. But when they are not, scientists are usually content to draw attention to a rough analogy and leave it at that. Since we have a discipline of fluid mechanics, but no discipline that studies flows and diffusions in a more abstract manner, this suggests that the properties and kinds in this case do not extend further than the realm of fluids or material substances. There is a special science dedicated to studying the flow of matter, but no broader scientific discipline or subdiscipline that studies flow or diffusion in a more general sense. However, it would be misguided to rest too much on this, since whether a scientific institution arises to dedicate itself to a set of kinds or properties would seem to be dependent on other factors than the extent and stability of the causal patterns that are held in common among the different phenomena.

Conversely, there are often interdisciplinary efforts that do not get institutionalized that devote themselves to such abstract similarities and succeed in unifying diverse phenomena, albeit in limited ways.

To summarize, the general answer as to whether a natural kind or real property is literally applicable to a new set of phenomena depends on whether such an extension results in generalizations that yield unifying explanations and novel predictions. This in turn depends on the extent and strength of causal relations in the new realm of phenomena. Extending the property of *viscosity* beyond the realm of material substances (e.g., to population genetics) would seem to be a borderline case in which neither are the parallels so strong as to result in a set of laws and explanations that constitute an independent field of study or subdiscipline that studies phenomena of flow and diffusion in general, nor are they so weak as to constitute a colorful analogy without any payoff in the form of the discovery of new generalizations or the discerning of new causal relationships. I will not try to pronounce with finality on this case because that would be to try to anticipate the progress of science. We have established that *viscosity* is a real property (at least if current science holds up); the only question is whether this is a property that pertains exclusively to fluid mechanics or whether it extends more broadly to domains such as population genetics.

3.6 LEVELS OF EXPLANATION AND CROSSCUTTING KINDS

In previous sections, I mounted a defense of special-science properties and kinds, which I argued are multiply realizable yet not reducible to lower-level properties and kinds, have causal powers, serve to unify diverse phenomena, and figure in empirical generalizations that are indistinguishable from many of the natural laws of the basic sciences. Occasionally, I have adopted the language of levels of description and explanation. I have also evinced a commitment to the existence of levels of *reality*, since I have defended the view that science tracks real causal relationships at each of these levels and that causal relations at the macrolevel can coexist with causal relations at the microlevel. Before concluding this defense of special-science properties and kinds, it is worth trying to get clearer on the significance of this talk of levels and attempting to make it less metaphorical; in the process, I will argue that the levels metaphor is misleading, partly because the relationship of multiple realization is not the only one that can obtain among levels.

There is a fairly mundane sense in which some phenomena of scientific interest differ in spatial dimensions from others, and, more importantly, some of these phenomena can only be captured at some spatial scales and not others. When it comes to an individual molecule or even of several molecules, there is no such thing as *viscosity*. The property only pertains to large collections of molecules, primarily in the liquid and gaseous phases of matter. The phenomenon of fluid flow does not even arise in the context of an individual molecule, and the diffusion of one fluid into another is only at issue when we focus not on single particles but rather on vast collections of them. Indeed, one of the fundamental assumptions of fluid mechanics, repeated in numerous textbooks, is that fluids are conceived as continuous media, not consisting of atoms or molecules, with continuous values for variables such as velocity and momentum. Properties like *viscosity* or *concentration* do not make sense unless we "zoom out" enough to be able to capture them at larger spatial dimensions. Meanwhile, other phenomena can only be considered if we "zoom in" enough to detect them at very small spatial dimensions. As Wimsatt (1994/2007, 207–208; original emphasis) observes: "[A] number of causal interactions characteristically become significant or insignificant together for things in a certain size range. *Size is thus a robust indicator for many other kinds of causal interactions.*"

A similar point can be made when it comes to temporal rather than spatial dimensions. Natural phenomena cannot be captured unless we observe or measure them at the appropriate temporal scale. The evolution by natural selection of many species of living organisms is simply not in evidence if we are observing things over the course of seconds, hours, or even years.[18] Equally, some elementary particles last for mere microseconds in a linear accelerator before decaying, and unless we find a means of detecting their existence over such a short time span, they will be invisible to us. That is not to say that *all* the eventual effects of such fleeting occurrences will be confined to such short time spans. There may well be an event that takes place over a period of a few seconds, such as a genetic mutation, which influences the course of evolution for an entire lineage of organisms for thousands of years. Since causal processes unfold differently in time, with some taking place in nanoseconds and others over the course of tens of billions of seconds, with all temporal scales in between, the patterns produced by these causal processes can only be

---

[18] This is not to deny that natural selection can operate over much shorter time scales, as in the microevolution of strains of bacteria that are resistant to a certain antibiotic compound.

captured if we adopt the right instrument, suitable for the right temporal scale, to make our observation. Though human beings cannot perceive events that unfold in nanoseconds, we have developed measuring devices that can, and though we cannot take in causal processes that take place over billions of seconds, we have instruments as well as indirect ways of gathering evidence to help us overcome that limitation.

It is worth remarking that while these two dimensions are relatively independent of one another, there appears to be some correlation between events that take place across extremely small distances and those that take place over very short periods of time, as well as those that take place across large spatial distances and over longer temporal durations. The reason for this seems reasonably clear. When one is dealing with macrospatial causal processes, the process as a whole involves the interactions of large numbers of particles or systems with multiple parts. Even though the properties and kinds involved in such a macroprocess may not be reducible to the properties and kinds of its parts, there is no denying that macrospatial processes are composed of these parts and that each instance of such a process involves a multitude of instances of microspatial processes. The completed causal process, whether it is the diffusion of one fluid into another, or the speciation of a population of biological organisms, or the transformation of a red giant star into a white dwarf, will supervene on many such processes, some occurring simultaneously and others in succession. Hence, it should be expected that macrospatial processes will typically be macrotemporal. With perfectly simple microspatial processes, by contrast, such as the decay of an atomic nucleus, the process is also microtemporal. The correlation is by no means perfect (and probably not capable of precise quantification) but it exists.[19]

"Levels" pertain not only to spatiotemporal dimensions, for there are descriptions of phenomena pitched at the same spatiotemporal level that pick out different *aspects* of these phenomena. Consider perhaps the simplest such case, drawn from nuclear physics (and mentioned in the previous chapter). Each kind of atomic nucleus (nuclide) can be classified according to atomic number (Z) or according to mass number (N). As we saw in the previous chapter, an atom of *lithium-8* ($Z = 3$, $N = 8$) can be classified with atoms belonging to other isotopes of *lithium* (e.g., *lithium-6* and *lithium-7*), which share an atomic number. But such an atom can also

---

[19] Compare Wimsatt (1994/2007, 216–217), who credits Simon (1962) with the insight that "processes at higher levels (with a few important exceptions) tend to take place at slower rates than processes at lower levels."

be classified with atoms belonging to the same isobar (e.g., *helium-8* and *beryllium-8*), which share a mass number. The former is a class of nuclides all of which contain 3 protons in their nuclei, while the latter is a class of nuclides all of which contain 8 nucleons in their nuclei. There are clearly important reasons for classifying all nuclides with atomic number 3 in the same class; indeed, that is the category commonly identified with the natural kind *lithium*. Is there a rationale for classifying all nuclides with mass number 8 in a single category? To be sure, isobars do not share as many properties as isotopes and they are not as strongly projectible. As I mentioned in section 2.5, both *lithium-8* and *helium-8* have a short half-life and they both decay by beta-minus decay; however, *beryllium-8* decays into two alpha particles. Most isobars have one beta-decay stable nuclide, which tends to be the nuclide with the greatest binding energy for that mass number. Other nuclides in that isobar tend not to be stable but exhibit radioactive decay. Beyond that, for any given isobar, one cannot make many generalizations that involve other projectible properties. Thus, it may be said that isobars are not good candidates for natural kinds, or they are at best weakly projectible natural kinds. But we can also group atoms of *lithium-8* together with *helium-8* in the category of *beta-minus decay nuclides*, where beta-minus decay is a process in which a neutron converts into a proton while emitting an electron and an electron neutrino. Radioactive nuclides of many elements exhibit this characteristic pattern of decay, which leads to a nuclide with the same mass number but a different atomic number (one less than the original) and a different charge. Such nuclides have genuine causal properties in common and a number of generalizations can be made about them. Since classification other than by atomic number is warranted for nuclides, this gives rise to crosscutting natural kinds, each system of kinds picking out a different aspect of the phenomena being classified. These classifications are pitched at the same spatiotemporal level and concern the same individuals but they pertain to different types of causal process that the relevant individuals can enter into (see Figure 3.1).

A similar case of crosscutting scientific categories in the (special) physical sciences comes from the classification of stars. Ruphy (2010, 1111) observes that astrophysicists involved in stellar classification are interested in several features of stars, such as temperature, density, and mass loss. Since these features are independent, stars can be cross-classified with respect to these features, none of which are primary or basic. Two stars can be classified in the same category with respect to one dimension but not according to another. Stellar classification by spectral types or temperature

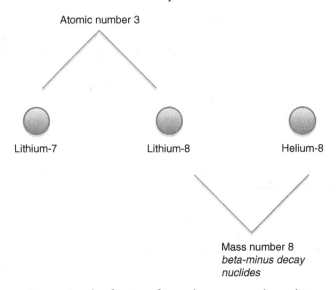

Figure 3.1. Crosscutting classifications of atoms by atomic number and mass number
(and pattern of decay)

utilizes the letters *O*, *B*, *A*, *F*, *G*, *K*, and *M*, to categorize stars from hottest
to coolest, but stars can also be classified in luminosity classes, using the
Roman numerals I to VII (sometimes supplemented with lowercase letters
to make finer distinctions). Consider Canopus (Alpha Carinae), which is a
type *F* star in luminosity class Ib; Procyon A (Alpha Canis Minoris A),
which is also a type *F* star in luminosity class IV; and Antares (Alpha
Scorpii), which is a type *M* star in luminosity class Ib. Procyon A and
Canopus are both type *F*, while Canopus and Antares are both in luminosity class Ib, but none of the categories in these two taxonomies contain
all three (see Figure 3.2). Are these both valid classifications that carve the
stellar domain into natural kinds? Astrophysicists use them to project other
properties of stars, in contrast with many other possible classification
schemes that are not projectible, such as classification into constellations
or according to distance from our solar system. However, Ruphy (2010,
1117) rejects the conclusion that stellar categories are natural kinds for three
reasons. First, she observes that stellar kinds do not have sharp boundaries,
since the dimensions along which stars are classified are continuous.
Second, she argues that crosscutting itself precludes thinking of stellar
kinds as natural kinds, particularly since there is no classification scheme
that divides stars most finely into "infimic species" (in other words, there is

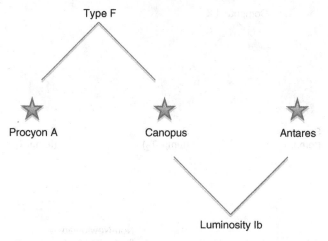

Figure 3.2. Crosscutting classifications of stars according to spectral type and luminosity

no bottom level in which there is a single, fine-grained system of classification that cannot be further subdivided). And thirdly, stars can migrate from one category to another as each star progresses through its life cycle. But it should be clear from this chapter and the previous one that all three features are attested to in the case of paradigmatic natural kinds: Some chemical compounds (e.g., isomeric compounds) and most biological species have no sharp divisions between them, there are no "infimic species" in the case of atomic nuclei (to divide them into nuclides by specifying both an atomic number and mass number is to effectively conjoin the two systems of classification, chemical and nuclear), and both elementary particles and atomic nuclei can be transformed from one category into another. Despite acknowledging that some of these features attach to other paradigmatic natural kinds, Ruphy considers that stellar kinds are not natural kinds in a realist sense but at best kinds that fulfill our epistemic needs. However, the argument that I have been making is that genuinely "epistemic kinds," which are projectible because they enter into real causal relations, are the only natural kinds there are.[20]

The same phenomenon of crosscutting natural kinds can be illustrated by the phenomena of fluid mechanics that I have been discussing in this

---

[20] Ruphy's relegation of stellar kinds to nonnatural kinds considerably weakens her ability to conclude that she has uncovered a pluralistic taxonomic scheme in the heart of the physical sciences (as opposed to the biological or social). But this is a conclusion that I heartily embrace, having also shown that it also holds for other domains in the physical sciences.

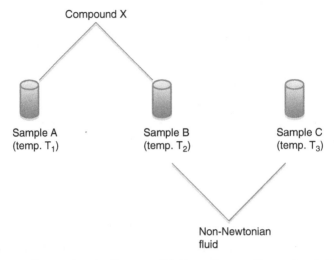

Figure 3.3. Crosscutting classifications of fluid samples according to chemical kinds and fluid mechanical kinds

chapter. Consider Newtonian fluids: Some substances have a Newtonian phase and a non-Newtonian one depending on temperature. Suppose that fluid sample A (chemical compound X at temperature T_1) is Newtonian, fluid sample B (chemical compound X at temperature T_2) is non-Newtonian, and fluid sample C (chemical compound Y at temperature T_3) is non-Newtonian. From the point of view of chemistry, A and B are both samples of compound X; from the point of view of fluid mechanics, A and C are both non-Newtonian fluids. But neither category, *compound X* and *non-Newtonian fluid*, contains all three fluid samples. Basic chemistry classifies two samples as belonging to the same kind due to molecular structure (not temperature), while fluid mechanics is primarily interested in the flow and diffusion of fluids, not their chemical composition as such (see Figure 3.3).

These instances of crosscutting categories of the same individuals (nuclides, stars, and samples of chemical compounds) show that the usual characterization of the relationship between levels as one of multiple realizability is not always apt. In the first two cases, the classification schemes are pitched roughly at the same spatiotemporal level, but in the third case the two levels are separated by several orders of magnitude in terms of spatial dimension. In this last case, what determines membership in the chemical kind is the molecular structure of the sample in question, while the determinant of its fluid mechanical kind is its macroproperties.[21] Multiple realizability posits a one-to-many mapping between higher levels and lower levels. But in this last case, there is a many-to-many mapping between the macro and micro. At first, this might seem utterly mysterious and metaphysically suspect. In section 3.2, I mentioned the widely accepted claim that the macro is supervenient on the micro, and that there can be no macrodifference without micro-difference. So how can it be that there are two samples of a single microkind, a chemical compound, that do not belong to the same macrokind, *Newtonian fluid*? Obviously, the difference lies in the temperature of the two samples of fluid (and temperature itself is multi-ply realizable in terms of microproperties). There is indeed a difference at the microlevel that accounts for the macrolevel difference in kind, thus dissolving the mystery. Generally, if the supervenience relation does not just pertain to the structural or even the intrinsic properties of microlevel phenomena (nor to macrolevel phenomena), there can be many-to-many relationships between microproperties and kinds on the one hand and macroproperties and kinds on the other. Supervenience between levels fails if it is taken to be local or intrinsic, but it is upheld as long as it is considered to be global and extrinsic (or relational).

The ubiquity of crosscutting taxonomies in different scientific domains suggests that something similar may obtain when it comes to psychological and neural states. On a common conception of the relationship between the psychological and the neural, there is multiple realization of the psychological in terms of the neural, leading to a one-to-many mapping between psychological states and neural states. This is the picture I assumed earlier in this chapter in responding to philosophers like Kim and Shapiro. But some considerations would tend to suggest a different picture, according to which psychological kinds may crosscut neural kinds rather than being multiply realizable by them. On this picture, person *A* could be in the same psychological state as person *B*, while person *B* is in the same neural state as person *C*, even though *A* and *C* are not in the same neural state (see Figure 3.4). How could this obtain? This may transpire, for example, if mental states are individuated extrinsically or relationally while neural states are individuated intrinsically. If externalism about the mental is correct, then the very same internal (neural) state of a person

---

[21] But bear in mind that the natural kind *water* may have a graded structure with focal and marginal members, or there may be no single natural kind *water* but several natural kinds with slightly different microstructures (see Hendry 2006; Needham 2000; Weisberg 2006).

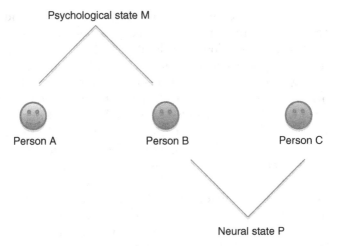

Figure 3.4. Crosscutting classifications of persons according to psychological kinds and neural kinds

could correspond to different psychological states depending on the context, external environment, or social community. If this situation obtains among some psychological and neural kinds, it would be similar to the situation that obtains among systems of kinds in the other special sciences (Khalidi 2005). Though there may be some instances in which one system of kinds is multiply realizable in terms of another, and other instances in which one system is reducible to another, there are other instances in which they comprise crosscutting systems. This constitutes yet another obstacle to considering special-science kinds to be equivalent to disjunctions of basic-science kinds, since the crosscutting relationship is many–many rather than one–many and precludes a disjunctive one.

Scientific disciplines and subdisciplines generally differ from one another along at least two dimensions, spatiotemporal scale and what I have been calling "aspect". In earlier work, I characterized this latter dimension as the *interest* with respect to which we investigate a certain phenomenon (Khalidi 1998a). Neither characterization is wholly apt and there may be no way of fully specifying this dimension without circularity; that is, without referring to the laws, causal processes, entities, properties, and kinds that are characteristic of that level. Generally, the aspectual dimension of levels is due to the fact that the same phenomenon can participate in orthogonal causal processes. Levels arise because the same set of entities, individuated in more or less the same way, can enter into causal relations and processes that are relatively independent of one another. This is what occurs with both nuclides and stars. In the case of nuclides, an individual atom can enter into (what can be loosely labeled) chemical and nuclear causal processes, respectively, in virtue of different properties possessed by each atom. In some cases, the difference in aspect is compounded by the difference in scale, giving rise to levels that differ in size and tempo as well as in salient causal processes; this is the case for the science of fluid mechanics, in relation to basic physics and chemistry. Hence, levels are always distinguished by aspect but they are sometimes distinguished by spatiotemporal scale as well.[22] However, I would argue that few if any levels are distinguished purely by spatiotemporal scale and that all of them are distinguished by characteristic causal relationships, as well as laws, properties, and natural kinds, rather than by peculiar observer-relative perspectives.

This can be considered the *raison d'être* of the special sciences and it is what grounds their claim to demarcate genuine aspects of reality as opposed to mere subjective perspectives on reality. The kinds and properties that they identify are not only objectively invisible at different spatiotemporal scales, their kinds and properties enter into causal processes that are largely independent of one another (like chemical reactions and nuclear decay patterns) and crosscut one another (like the temperature and luminosity of stars). Each of these taxonomic systems classifies phenomena on the basis of genuine causal processes rather than according to the subjective perspectives of certain observers. Observers track these processes, to be sure, but that does not mean that they are observer-relative. To use a term first introduced by Dennett (1991), these causal processes are "real patterns"; indeed, one could talk of *real causal patterns*.[23]

Since I have been arguing that the special sciences do not always relate to one another or to more basic sciences simply as macro to micro, talk of "levels" is better replaced by talk of "domains." The "levels" metaphor is

---

[22] In this respect, I differ with Wimsatt (1994/2007, 227), who distinguishes "levels of organization" from "perspectives," on the grounds that levels are related to one another compositionally while perspectives are differentiated by "a set of variables that are used to characterize systems or to partition objects into parts, which together give a systematic account of a domain of phenomena, and are peculiarly salient to an observer or class of observers because of the characteristic ways in which those observers interact causally with the system or systems in question."

[23] In a similar vein, Ladyman and Ross (2007, 200–204) have discussed the "scale relativity of ontology," and they have also endorsed cross-classification in the special sciences. But they do not appear to distinguish the spatiotemporal and the aspectual dimensions of scientific disciplines as I have done, nor do they give clear illustrations of crosscutting kinds apart from different concepts of biological species (which can coexist according to the pluralist position concerning species).

misleading because it implies that the relationship between the phenomena at successive levels is hierarchical. It is also misleading because it suggests that phenomena at higher levels relate as one-to-one, or perhaps one-to-many, to those at lower levels, whereas I have argued that the relationship is one of crosscutting or many-to-many. Scientific domains are not hierarchical because although they can be ranked along the spatiotemporal dimension, there is no sense in ranking them along the aspectual dimension.[24] Does specification of these two dimensions, the spatiotemporal and aspectual, uniquely specify a domain? For any given domain, D, there may be no way of specifying necessary and sufficient conditions to single out the purview of D in terms not derived from the theory or theories that apply to D, since even the individuals identified in a domain are generally picked out against the background of the theory or theories prevalent in that domain. In some cases, the individuals classified according to crosscutting systems of kinds may not be exactly coincident: Mass number applies to the atomic nucleus, while atomic number pertains to the whole atom (including electron orbitals); mental states are states of persons, while neural states are states of the brain. Finally, even though I have been speaking as though each domain corresponds to a single special science, things are seldom so neat in practice, since the same special science can comprise more than one domain (just as stellar astronomy classifies stars into crosscutting categories). Be that as it may, the relationship between kinds drawn from distinct domains as well as that between the kinds within a single discipline is often one of crosscutting rather than multiple realization of the macro in terms of the micro.

3.7 CONCLUSION

The picture of the special sciences that emerges from this discussion is of disciplines or subdisciplines that aim to identify properties and kinds on the basis of causal relations detected in proprietary domains. Domains are both spatiotemporal and aspectual, but they may not be capable of being individuated noncircularly without recourse to the theories, properties, and kinds that occur in those domains. The causal relations in each domain can be summarized in the form of laws and generalizations that relate these properties and kinds in various ways, both quantitative and qualitative, and enable us to project them to new situations and

---

[24] Ladyman and Ross (2007, 54) likewise eschew talk of "levels" on the grounds that scientific disciplines are not arranged hierarchically.

phenomena. The kinds of the special sciences are related to the kinds of the more basic sciences and other special sciences in complex and attenuated ways. First, differences in spatiotemporal scale mean that special-science properties and kinds are often simply not identifiable at the more basic or microlevel and that they are only applicable to macrolevel entities. Second, the properties and kinds of the special sciences are not merely disjunctive but serve to unify diverse phenomena under causal laws and generalizations that apply to phenomena that may not have been part of their original purview. Third, the entities individuated in a special-science domain are not always coincident with or mere aggregates of those in the domain of any one basic science, since they are sometimes individuated extrinsically or relationally. Fourth, special-science properties and kinds are not just multiply realizable in the properties and kinds of other sciences (whether special or basic), since they often crosscut them, classifying the same or nearly coincident individuals in partially overlapping natural kinds. This last point does not violate metaphysical supervenience, provided that supervenience is understood globally and relationally.

At least some of these conclusions agree broadly with those of other philosophers who have taken a realist position on the ontology of the special sciences and the reality of special-science kinds and properties. In addition to those philosophers whose views I have referred to in this chapter and the previous one, Dennett's widely known accounts of "real patterns" and the "scale relativity of ontology" (Dennett 1991; cf. Ladyman and Ross 2007) are clear precursors. Though Dennett does not emphasize causation, it is compatible with his account, to the point that one could talk of "real causal patterns" without doing violence to his position. My defense of special-science kinds also partially accords with others who have argued for natural kinds and natural laws or empirical generalizations in the special sciences (Fodor 1997; Kistler 1999; Kornblith 1993). To these views, I have tried to add a more precise characterization of scientific *domains*, distinguishing their spatiotemporal dimension from their aspectual dimension. I have also attempted to put forward a more concrete understanding of the phenomenon of crosscutting properties and kinds, illustrating it through various examples. This account of natural kinds in the special sciences, which relates them to real causal patterns in cross-cutting domains, develops further the naturalist position that I defended in Chapter 2. I concluded that chapter by arguing that the projectibility of natural kinds reflects their *causal* nature, but I also argued that the causal roles of natural kinds are more variegated and diverse than envisaged in Boyd's HPC theory of natural kinds, and this has led us to what Craver

(2009) calls a "simple causal theory" of natural kinds. We now have a better understanding of the relevant causal relations and processes, how they differ in distinct domains, what enables causal processes in micro- and macrolevel domains to coexist peaceably, and how some of these causal processes can be discerned in new phenomena outside their original areas of application, thereby unifying apparently disparate phenomena. This emphasis on causality is also in line with the proviso that I introduced in Chapter 2, which gives primacy to categories introduced for *epistemic purposes* as opposed to other purposes. The idea that scientific knowledge (*episteme*) is of causes and causal relations is at least as old as Aristotle, but it is also very much alive in the scientific enterprise to this day. This is not *epistemicism* but *naturalism*, for if metaphysics is not a purely a priori enterprise, then it must be informed by our surest knowledge-gathering endeavors and cannot proceed independently of our epistemic practices.

CHAPTER 4

Kinds in the biological and social sciences

4.1 BIOLOGICAL AND SOCIAL KINDS

In the previous chapter, I defended the existence of natural kinds in the special sciences, focusing primarily on kinds drawn from those special sciences that are not biological or social in nature. These two classes of sciences, biological and social, have features that have been thought to set them apart from other special sciences, and may also be thought to preclude their categories from corresponding to natural kinds. But each of these two classes of sciences, biological and social, may also be thought to raise their own peculiar questions, so why should we lump them together? I will argue that at least some of the features that pertain to the biological sciences pertain to the social sciences as well, and that it is informative to consider these features as they occur in both, but I will also be considering separately those issues that pertain only to one or the other. Moreover, the dividing line between biological and social sciences is not a sharp one, and psychology and the cognitive sciences are commonly thought to lie somewhere in the area of overlap.[1] The kinds deriving from the biological and social sciences are often referred to as biological and social kinds, respectively, but the philosophical literature sometimes also mentions "human kinds," which I take to be roughly coextensive with social kinds, so I shall use these two expressions more or less interchangeably. It is true that there can be human kinds that are not obviously social (e.g., kinds of diseases specific to human beings), and there may also be social kinds that are not human kinds (e.g., *dominant male*, as applied to macaque monkeys).[2] But both these types of kinds can be considered biological kinds, so treating biological and social kinds together has the added advantage of not neglecting nonsocial human kinds and nonhuman

[1] For notable arguments to the effect that there are important similarities between biological and social kinds, see Ereshefsky (2004) and Dupré (2004).
[2] The latter claim is made by Ereshefsky (2004).

125

social kinds. Moreover, the existence of these two classes of kinds further confirms the porous nature of the boundary between biological and social kinds.

This chapter will be organized around a number of specific arguments that have been put forward by various philosophers for thinking that biological or social kinds are fundamentally different from kinds in other domains. I will argue that only one of these arguments poses a serious threat to considering biological or social kinds to be natural kinds in the sense that I have advocated in the previous two chapters, but I will also point out that that argument does not threaten all biological and social kinds. While some of these arguments apply to both biological and social kinds, others apply primarily to social kinds (though they may also apply to some biological kinds, particularly those human biological kinds that are arguably not social).

4.2 SELECTED KINDS AND DESIGNED KINDS

Some philosophers who have taken a dim view of special-science kinds in general make an exception for those kinds that result from a process of natural selection or those that are deliberately selected or explicitly designed. While they may reject other special-science kinds, they accept some kinds in the biological (and perhaps social) sciences on the grounds that they are the result of a selection process. This puts me in the position of agreeing with them on the naturalness of some of these kinds, but for different reasons. For the sake of clarity, I will try to explain why I do not think that natural selection or explicit design is the reason that there can be kinds in the biological and social sciences, though there can be such kinds nevertheless.

Anticipating some of the arguments of Kim and Shapiro discussed in sections 3.2 and 3.3, Papineau (1985, 2009, 2010) has put forward related reasons for rejecting autonomous special-science kinds and properties that are not reducible to the kinds and properties of the basic sciences. Papineau's objection to the existence of such kinds can be put in terms of a puzzle. Let us suppose that a special-science kind K is multiply realized by some (possibly open-ended) set of basic-science kinds, $\{L_1, L_2, \ldots\}$, and that K is linked by means of a law or empirical generalization to some other special-science kind K^*, which is itself multiply realized by a set of variegated basic-science kinds, $\{L^*_1, L^*_2, \ldots\}$. Papineau (2010, 180) wonders how is it that, if the realizations of K are all so different from the point of view of the basic sciences, they all give rise to states that realize the very same special-science kind K^*? He thinks that it is nothing short of a miracle that they all somehow converge on a single special-science kind, and he illustrates this point by means of a hypothetical example. Imagine that we found that all *people who eat reheated Brussels sprouts* (K) are *people who suffer from inflamed knees* (K^*), but that in one type of case the reheated Brussels sprouts harbor a *virus* (L_1) that *infects the knees* (L^*_1), whereas in another they contain a *high level of uric acid* (L_2) that leads to *gouty attacks* (L^*_2), and so on. Papineau (2010, 181) comments on this state of affairs as follows:

This story doesn't hang together. It beggars belief that reheated Brussels sprouts should always give rise to inflamed knees, yet the physical process that mediates this should be different in every case. Surely either there is some further feature of the sprouts that can explain why they all yield the same result, or we were mistaken in thinking that there was a genuine pattern in the first place, as opposed to a curious coincidence in our initial sample of cases.

But Papineau claims that when it comes to some cases in the biological sciences, selective causal processes can give rise to autonomous kinds. To show how this can arise, he begins by considering a device that has been explicitly designed by humans to perform a certain function – namely, a thermostat. Different thermostats rely on different mechanisms to regulate temperature by breaking a circuit that turns off a heating device when the temperature has increased to a certain level. "If there is no uniform physical explanation for this commonality," Papineau (2010, 186) asks, "is it not a mystery that all the divergent effects of temperature increases should converge on this single effect [i.e., breaking the circuit]?" In this instance, the question is rhetorical because Papineau thinks there is an obvious solution to the mystery – namely, the existence of human designers who have arranged it so that these devices converge on the effect of switching off the device when a certain temperature has been reached. When there is intelligent *selection* for an effect, it is clear why divergent causes can converge on a single effect. Moreover, since natural selection is a process that yields designed mechanisms (though there is no intelligent designer), Papineau thinks that the same holds for the products of natural selection. For instance, he allows that the kind *pain* is "variably physically realized across different life forms," but he says that the reason that these different causes converge on the same effect in this case is that natural selection has effectively designed them thus. Papineau (2010, 186) writes: "Natural selection favors organisms that have some mechanism that mediates between bodily damage and the avoidance thereof. It doesn't care too much about how this is done." This rehabilitates at least some biological, psychological, and (perhaps) social kinds, and shows how they can be both multiply realized but not merely disjunctive, precisely because they have been selected for a certain effect, either naturally or artificially. It is not that they all mysteriously converge on a certain effect but that they have been *selected* for their ability to do so.

Before going on to see why Papineau's rehabilitation of such kinds does not amount to a full-blown endorsement, I need to say why I think that his defense is not as it should be. In my own account of special-science properties and kinds in Chapter 3, I argued that these kinds should be considered to be *aspectual* in nature. *Viscosity* is a legitimate scientific property despite the fact that it is multiply realized in gases and liquids because it captures an aspect of the dynamics of fluids and characterizes the behavior of a variety of fluids in a range of causal processes, enabling us to explain the phenomena of fluid flow because this property figures in causal laws. Similarly, *Newtonian fluid* is a natural kind because it includes a set of different liquids and gases that all share a complex property and enter into the same causal laws. To try to demystify the puzzle that Papineau invokes, it is not that divergent causes somehow conspire to converge on a single type of effect, but rather that in considering these phenomena we are only concerned with a subset of their causal properties, and these properties are not detectable at the level of basic microphysics. Suppose someone were to ask, how is it that gaseous oxygen, liquid water, and liquid benzene all somehow realize the kind *Newtonian fluid*? And by what bizarre coincidence do they all, despite their very different microstructures, obey the same causal laws, as in the Navier–Stokes equation? The answer provided in the previous chapter is that they do so because when we focus on a certain subset of their causal properties (those relating to fluid flow) we find that they participate in the same causal processes and give rise to similar effects at the macrolevel. Meanwhile, many of their divergent microproperties can be ignored (e.g., chemical properties, nuclear properties, and so on). Despite the fact that the microphysical processes that underwrite these causal properties are different in these different cases (in particular, the nature of the chemical bonds are widely divergent), they share a rich set of properties at the macrolevel and participate in many of the same causal processes.

It may be thought that it is crucial in the case of fluid mechanics that the macroproperties and kinds are *physical* properties and kinds, and that this is what gives them their causal unity. But that is surely not necessary to the example and does not differentiate these broadly functional kinds from other special-science kinds. The properties and kinds that we identify in biology, as in other special sciences, participate in robust causal patterns that recur in nature in various contexts. Tracking them requires ignoring certain dissimilarities or focusing on certain aspects of causal reality to the exclusion of others from a particular spatiotemporal perspective. Hence, I would argue that Papineau's approach to rehabilitating biological kinds is mistaken. Biological kinds are natural kinds when they are implicated in real causal patterns in a certain domain, as explained in the previous chapter. If they are, then they will be vindicated as natural kinds. Selection mechanisms do not provide the underlying reason for the unity of these kinds. If members of these kinds or instances of these properties did not have the ability to produce a common effect, then they would not have been selected to do so in the first place. In the case of a thermostat, the common causal properties of a bimetallic strip and a thermocouple are what allow them to produce the desired thermostatic effect, whether we select them for this purpose or not. We select them after having discovered their common properties, ignoring various other properties that they might have. Natural selection does the same but without design.

To return to Papineau's limited defense of biological kinds: After he allows that selected or designed kinds are allowable even though other special-science kinds are not (unless they are *reducible* to physical kinds), he goes on to demote such kinds for other reasons. Using the example of the multiply realized category *pain*, Papineau (2010, 188–189) writes that even though we can make some limited cross-species generalizations about pain (e.g., pain leads to damage avoidance) these generalizations are rather limited in nature. By contrast, the species-specific kind *pain* (e.g., *human pain*) enters into a rich causal network of laws due to the fact that the physical realizer of *pain* is more or less the same in members of the same species.[3] Thus, he holds that special-science properties and kinds, particularly biological and social kinds, do not themselves enter into a "complex of laws" though their realizers do in specific species or systems. Without denying that there are such laws that apply only to the local realizers of special-science properties and kinds, I have already argued (above, and in Chapter 3) that there are also nontrivial causal laws that apply across a range of phenomena in the special sciences, not just those that appeal to specific microstructures. Moreover, even though we gain specificity when we concentrate on the laws that apply to a single subsystem (e.g., pain in a

[3] As we saw in Chapter 3, both Kim and Shapiro make similar claims about nonreducible special-science kinds.

particular species), we lose the ability to *unify* a range of phenomena under the same general laws. What remains to be shown in more detail, of course, is that the same considerations apply to at least some of the kinds in the biological and social sciences – namely, that they, too, feature in causal laws that serve to unify otherwise diverse phenomena. When it comes to any specific biological or social kind (e.g., *pain*), we cannot determine whether or not it is a natural kind without a closer examination of the case in question. I will not attempt to do so in this chapter, but will postpone a more detailed consideration of this question to Chapter 5, when I discuss specific case studies. For the moment, my aim is to overcome a few principled obstacles to the inclusion of biological and social kinds among the natural kinds.

4.3 ETIOLOGICAL KINDS

One challenge to the existence of biological and social kinds has to do with the fact that they are frequently individuated etiologically.[4] To focus again on biological kinds, members of the same kind are often grouped together not by virtue of the fact that they possess some of the same synchronic causal properties, but because of the fact that they share a causal history; conversely, individuals that share many synchronic causal properties but do not share a causal history are often placed in different categories. Although some biological taxonomists accord primary importance concerning species to the synchronic features of organisms (principally members of the phenetic school of taxonomy), the dominant view is to set more store by their diachronic features – namely, phylogeny, descent, or causal history. Moreover, systematists do this to different extents. As mentioned in section 2.5, cladists exhibit perhaps the strictest adherence to classification by descent among biological taxonomists, though others accord descent at least some role in taxonomy.[5]

The reason for the emphasis on causal history is not hard to ascertain in biology, since the theory of evolution is the central scientific theory in the biological sciences and evolutionary theory explains the diversity of life with reference to a history of natural selection. The difference between

---

[4] In this section, I will not discuss social kinds, since I think the claim that some social kinds are etiological is uncontroversial. Moreover, the main question to be addressed is whether etiological kinds can be natural kinds, so biological examples will suffice. Compare Dupré (2004, 896): "I suggest that … like human kinds, biological kinds must often be understood historically."

[5] There are far more species concepts in current use among biologists than can be discussed here; for example, Hey (2001) enumerates twenty-four species concepts.

classifying by (synchronic) causal properties and by (diachronic) causal history does not show up very prominently in the classification of individual organisms into species since these two features tend to coincide at the level of individual organisms (those organisms with a shared causal history tend to share many causal properties in common, and vice versa). But it does loom large when it comes to the classification of species and other taxa into higher taxa (genus, family, order, and so on). Here, classification often sets aside similarities in causal properties (even nonsuperficial differences) in order to follow proximity of descent, or it tolerates significant differences in causal properties in order to track descent from a common ancestor. That is why the *red fox* is classified more closely with the *raccoon dog* than with the *gray fox* despite the similarities between red and gray foxes, and why the *raccoon dog* is not classified with the *raccoon* despite their resemblance. Convergent evolution can yield species of organisms with many of the same synchronic causal properties, such as the *European mole* and the *marsupial mole*, yet many biologists would not classify such groups in a common taxon on the basis of these properties because they do not share a recent common ancestor. The same principle can be seen at work in the identification of *homologies* among phenotypic traits in biological organisms. Some traits or features of organisms are classified together in different species of organisms, despite their differences in synchronic causal properties, because of the fact that they share a common ancestry. Thus, the radius bone in the leg of a mammal plays a very different causal role than the radius bone in the wing of a bird, yet both are descendants of the same bone in a vertebrate ancestor and are classified as homologous organs by evolutionary biologists.

These examples from biology provide a contrast between two different modes of classification. One of them categorizes individuals on the basis of their synchronic causal properties, or their powers to effect causal changes or initiate causal processes in the present or future. The other categorizes them based on their having been generated by the same or similar causal processes, or by virtue of having a similar diachronic causal trajectory, regardless of their synchronic causal powers. Each mode of classification attends to different types of properties that individuals may have, and it is one instance of the contrast between attending to intrinsic and extrinsic properties (which was first discussed in section 1.6, where I argued for the defensibility of natural kinds associated with extrinsic properties). Though etiological properties are not the only type of extrinsic properties, they constitute a prominent example of such properties. I have provided a reason for expecting that etiological properties and etiological kinds will be prevalent in the biological sciences, given the importance of evolutionary theory for understanding the biological realm, and have also provided a few examples of such kinds. But it remains to be seen whether all biological kinds are etiological, whether etiological kinds are confined to the biological (and social) sciences, and, more importantly, whether they are disqualified from being genuine natural kinds. In the remainder of this section, I will tackle these questions in turn.

Before we turn to the questions of whether all biological kinds are etiological and whether etiological kinds can be found outside the biological realm, it will be helpful to consider the views of another philosopher who has discussed etiological kinds in biology. Like Papineau, Bird (2008) denies the causal efficacy of special-science kinds and properties, but he makes an exception for biological kinds and properties precisely because in many cases they share a causal history. Bird notes rightly that many phenotypic features of organisms are considered members of the same kind because they share a causal history or because they result from similar selection processes. For example, the eye of a bee and the eye of a cat are different in terms of their causal powers but they have evolved as a result of similar selection pressures, and that justifies considering them to be members of the same natural kind. Bird (2008, 169) considers a cat and a bee that both use their sense of vision to navigate around a tree:

The immediate causes of vision in that bee and that cat are different – the differing genotypes of the two creatures. But the more distal cause, which is the same as the general cause of the presence of vision in bees and cats, is much the same. It is the selective advantage that vision gives creatures that possess it over those that do not.

Here, Bird uses the *eye* as an example of a natural kind that is individuated by causal history rather than causal power; that is, as the effect of similar causes rather than the cause of similar effects. But other philosophers (as well as many biologists) have regarded it as an example of the very opposite phenomenon – namely, a case of convergent evolution, in which *different* causal histories converge on organs with highly *similar* causal powers (much like the European mole and Australian "mole"). On that view, what would justify classifying both the bee eye and cat eye as members of the natural kind *eye* is (at least in part) the causal power to equip the organism with the wherewithal to perceive electromagnetic waves in a certain range and produce a two-dimensional image on a retina. When it comes to phenotypic traits, traits classified together on the basis of shared causal history are homologous, while those classified together on the basis of shared causal powers are *analogous*. It is common to regard the eyes of vertebrates and arthropods, or the dorsal fins of sharks and dolphins, or the thumbs of pandas and humans, as *analogies* rather than homologies (like the radius bone in the leg of a mammal and that in the wing of a bird). Despite broad similarities in selection pressures, it would be wrong to conclude that the eyes of vertebrates and those of insects have the *same* "distal cause" or causal history, as Bird does. Moreover, I would maintain with Weiskopf (2011) that the end products are dissimilar enough to qualify as multiply realized, but they also have enough in common in terms of synchronic causal powers to qualify as members of the same natural kind (as did some of the special-science kinds considered in Chapter 3).[6] We have already seen other examples in biology wherein individuals are grouped into etiological kinds and classification by causal history diverges from classification by causal properties, notably in the case of homologies. This example also (inadvertently) shows that that is not the only mode of classification in biology, since causal powers are also used for the classification of some natural kinds – for example, analogous traits. Hence, there can be classification by causal power as well as by causal history in biology.

There are other instances in science, not just in the biological sciences, where scientists classify according to etiology. Planetary geologists characterize a *meteorite* as a natural object on the surface of a celestial body that has come from elsewhere in space. Two rocks that are identical in material constitution or causal powers, one of which was part of a planet when it originally formed, while another originated elsewhere and collided with the planet at a later stage, may be classified differently by space scientists. In addition, geologists classify terrestrial rocks as *igneous, sedimentary*, and *metamorphic*, on the basis of the causal processes that gave rise to them. Igneous rocks are formed when molten magma cools, sedimentary rocks by a process of deposit and compaction, and metamorphic rocks as a result of changes in temperature and pressure. In this case, the rocks in each

---

[6] Some opponents of multiple realizability have argued that the eyes of vertebrates and those of other kinds of organisms do not have enough in common in terms of causal power to justify classifying them as members of the same natural kind. This is the conclusion of Shapiro (2004, 94–103), who argues here, as in other cases discussed in Chapter 3, that the eye has an analytic functional definition across diverse kinds of organisms rather than being a rich source of inductive inference, and hence is not a natural kind (cf. *jade*). Meanwhile, within vertebrates, for example, there are severe physical constraints on the properties that a light-sensitive organ must obey in order to form a two-dimensional retinal image, so the vertebrate eye is a singly realized natural kind. In response, Weiskopf (2011) argues convincingly, in my opinion, that arthropod eyes and vertebrate eyes are multiply realized members of the same kind, on the basis of their shared causal properties.

category tend also to have certain synchronic features in common, but they are classified at least partly on the basis of etiology. Yet another example of an etiological kind, this one from cosmology, would be the *cosmic microwave background radiation* (*CMBR*), which is thought to be left over from the big bang (and is one of the main sources of evidence for its occurrence). This kind of radiation, which everywhere has the same frequency, is characterized on the basis of its origin or the causal process that gave rise to it. Electromagnetic radiation with that very frequency would not be classified as *CMBR* unless it had originated from the same source.

Having provided some reason to think that there are both etiological and nonetiological kinds in biology and that there are also examples of etiological kinds outside of biology (e.g., in planetary science, geology, and cosmology), it remains to consider whether etiological kinds can be natural kinds. Given the account of natural kinds that I have elaborated so far in this book, does it follow that classification schemes based on etiology or causal history do not correspond to natural kinds? At first sight, it may seem so. Projectible properties are ones whose instantiation leads reliably to the instantiation of other properties, and this is generally a matter of causal power. However, since causal power and causal history can come apart, classification by causal history (when causal powers do not coincide) would seem to go against the thrust of the account of natural kinds that I have been defending. Hence, if at least some classification in the biological (as well as the cognitive[7] and social) sciences is by causal history or etiology, it may seem to follow that etiological kinds are not natural kinds.

But I would argue that this reaction is too quick. In previous chapters, I identified natural kinds with those kinds that are associated with projectible properties, are featured in empirical generalizations, and, more fundamentally, participate in real causal patterns. A liberal understanding of the features that I have associated with natural kinds would not dismiss all etiological kinds as nonnatural kinds. To start with projectibility, although all radius bones do not share in a large number of synchronic causal properties (not shared with members of some larger category, like vertebrate bones), they do share a causal history. Once a bone in a vertebrate has been identified as a radius bone, we can make inferences about the process by which this bone evolved, and hence we can say something about

---

[7] For an argument to the effect that neuroscience sometimes classifies by etiology or causal history rather than causal power, see Garson (2011). In Chapter 5, further examples will be provided of etiological kinds.

its causal history. To put it differently, retrodiction is also a type of projection. Similarly, we can make generalizations about the processes that led to the generation of members of etiological kinds like igneous rocks. When igneous rocks are discovered in different geographical areas, they often provide information about the composition of the Earth's mantle, since the molten magma typically originates there. Additionally, even though an etiological kind may constitute the endpoint in a causal process, it participates nonetheless in a real causal pattern. A complete causal account of the universe should preserve differences in causal outcomes as well as differences in the causal processes that led to those outcomes. Some of the examples mentioned of etiological kinds demonstrate that science is concerned not just with understanding the synchronic causal powers of individuals or entities but also with the causal processes that led to them being the way they are even when their current causal powers are identical. If real causal patterns are what we are interested in tracking, there is no reason to focus exclusively on the synchronic ones as opposed to the diachronic ones. When there is a mismatch between causal history and causal power, specifically when a shared causal history gives rise to individuals with different causal powers, it may be warranted to classify individuals as members of the same kind according to a diachronic taxonomic scheme, while at the same time classifying them as members of different kinds according to a synchronic taxonomic scheme. Therefore, from the account of natural kinds that I have been proposing, etiological kinds should not be disqualified.

Having defended the position that etiological kinds can be natural kinds, it is worth stating that etiological kinds do seem to constitute a genuinely different type of natural kind than the other kinds we have been considering. The properties associated with them do not generally lead to the causal generation of a series of other properties or initiate a sequence of other properties. Unlike eyes, radius bones do not share a number of synchronic causal properties that in turn lead to the instantiation of other causal properties. It should also be emphasized that causal kinds and etiological kinds are not mutually exclusive categories. As suggested already, biological species may be the pre-eminent example of natural kinds that are endowed with both characteristics and are capable of being individuated in terms of either, with some taxonomists (e.g., pheneticists) favoring synchronic causal properties and others (e.g., cladists) favoring diachronic causal properties. Though there are occasional divergences among causal power and etiology, and hence in the way that these taxonomists individuate species, the different taxonomic methods often do not diverge in practice. The reason is that the historical causal processes that give rise to biological species tend to reinforce uniformity of causal power, with some significant exceptions (e.g., polymorphism).

In previous sections, I argued that there can be more than one taxonomic scheme that classifies the same individuals differently for different purposes and that these schemes can crosscut one another (see sections 2.4 and 3.5). This would seem to be one instance of that phenomenon. A taxonomic scheme that classifies individuals by causal powers may well crosscut a scheme that classifies those same individuals by causal history, as with analogies and homologies concerning phenotypic traits or features. Since crosscutting classification schemes can coexist comfortably in science, if scientific disciplines sometimes classify individuals on the basis of their causal history, this should pose no obstacle to these classifications being natural kinds.

4.4 HISTORICAL KINDS OR COPIED KINDS

There is a distinction that is closely related to the one discussed in the previous section, between causal kinds and etiological kinds, which has been discussed extensively by Millikan (1984, 1999, 2000, 2005). Millikan observes that the members of many kinds in the biological sciences and social sciences (as well, perhaps, as some artifactual kinds and nonscientific kinds) are identical or similar to each other mainly because of a copying process rather than as a consequence of laws of nature. Atoms of *lithium-7* are identical and belong to the same natural kind because there are laws of nature whose consequence it is that three protons and four neutrons will form a stable nucleus due to the operation of the strong and weak nuclear forces, and further that this nucleus will be that of a stable atom also including three electrons, due to electromagnetic forces. But members of the kind *tiger* are similar to one another because they are descended from the same ancestor, not because it is a law of nature that organisms with (roughly) this genotype and phenotype are stable or capable of being created. Millikan (2005, 307–308) calls the former categories "eternal kinds" and the latter "historical kinds" or "copied kinds," and she characterizes them as follows:

A "historical kind" (like dogs, for example) is a collection of individuals scattered over a definite spatiotemporal area; the individuals are causally related to one another in such a manner that each is likely to be similar to the next in a variety of aspects. There are two most obvious sorts of things that cause members of a historical kind to be like one another. First, something akin to reproduction or copying has been going on, all the various individuals having been produced from one another or from the same models. Second, various members have been produced by, in, or in response to, the very same ongoing historical environment – for example, in response to the presence of members of *other* ongoing historical kinds. Third – an ubiquitous causal factor often supporting the first factor – some "function" is served by members of the kind, where "function" is understood roughly in the biological sense as an effect raising the probability that its cause will be reproduced. It is typical for several of these kinds of causes to be combined.

The first feature of copied kinds that she points out (copying or reproduction) is the primary one, while the other two features (historical environment and function) appear to be supplementary features that often serve to reinforce the first. Among other historical or copied kinds that Millikan (1999, 56–57) mentions are *McDonald's restaurants, doctors, 1969 Plymouth Valiants*, and *Californians*; in the same vein, Elder (2004) adds *screwdrivers, stickleback mating displays*, and *Eames 1957 desk chairs*.[8] It is worth emphasizing that Millikan's distinction between eternal and historical kinds does not coincide neatly with the distinction between causal and etiological kinds discussed in the previous section. Millikan is drawing attention to the existence of kinds, members of which are similar or identical in causal powers ("like one another") as a result of a copying process. Generally, etiological kinds, which have similar or identical causal histories, need not have similar or identical causal powers. Hence, it would seem as though Millikan's copied or historical kinds are examples of kinds that are both causal and etiological: Their members are classified on the basis of synchronic causal properties *as well as* a shared etiology (a copying process). Moreover, they are a particular type of etiological kind – namely, one in which the causal process that leads to their instantiation is a copying process.

It is not Millikan's intention to argue that only eternal kinds are natural kinds and that historical or copied kinds cannot constitute natural kinds.[9] Indeed, she emphasizes, for example, that copied kinds can feature in

---

[8] Note that when it comes to at least some biological and social kinds, the copying process is also guided by human design. It is significant that Millikan mentions the kind *dog* in this connection, which is the result of artificial rather than natural selection. Here, the copying process is one guided largely by human artifice. Humans are constrained in what kinds of organisms they can artificially select (or genetically engineer), but these organisms do not arise as a result of a simple copying process.

[9] Millikan (2000, 16) speaks of "real kinds" rather than "natural kinds" because she thinks that natural kinds are commonly thought to be *definable* by a set of properties and their properties are merely said to be *correlated* rather than shared due to a "real ground in nature". But neither condition applies to the term "natural kind" as I have been using it, as explained in previous chapters. Also, Millikan appears to use "historical kind" and "copied kind" interchangeably, but I will tend to use the latter to avoid confusion between historical kinds and the etiological kinds discussed in the previous section.

empirical generalizations, though not exceptionless ones. However, it seems to me that she has drawn attention to an intriguing distinction between two kinds of kinds and one that might be used to try to drive a wedge between physical and chemical kinds on the one hand (whether special or not) and biological and social kinds on the other. Accordingly, my aim in this section will be to try to answer three questions. First, does the distinction between eternal and copied (or historical) kinds correspond neatly to the distinction between the physical and chemical sciences (on the one hand) and the biological and social sciences (on the other)? Second, how deep and principled is the distinction between the two kinds of kinds? Third, can the distinction be used (*contra* Millikan) to argue that only eternal kinds are natural kinds and that copied kinds cannot be natural kinds?

To tackle the first question as to whether the distinction coincides with that between the physical-chemical and the biological-social, it bears repeating that it cannot be the case that *all* biological kinds are copied kinds, since we have already seen that some biological kinds are not even etiological kinds and since copied kinds must at least be partly etiological. It follows that not all biological kinds are copied kinds. The eyes of bees and birds are not descended from each other or from a common ancestor, so they are decidedly not copied; that is what makes them examples of convergent evolution. But are biological (and social) kinds the only copied kinds, or can there be copied kinds outside these realms? When one considers the paradigmatic physical-chemical kinds, such as elementary particles, chemical elements, or chemical compounds, and even special-science physical or chemical kinds such as *gases*, *Newtonian fluids*, and *stars*, it may seem obvious that members of these kinds do not resemble one another due to some copying process. Atoms of *lithium-7* are the way they are due to the operation of laws of nature that dictate the limits of stability when it comes to atomic nuclei. Meanwhile, many biological and social kinds, such as members of a species and practitioners of a profession, have the properties that they do as a result of copying processes (e.g., sexual or asexual reproduction, learning, imitation) that replicate certain features occurring in other members of the kind. But before we conclude that all physical-chemical kinds are eternal kinds and that copied kinds can only exist in the biological and social realms, there are several complications to be borne in mind. First, consider the chemical substance responsible for most of the copying that takes place in the biological realm, DNA. Identical strands of DNA are such as a result of a chemical reaction involving other members of their kind (and other kinds such as the enzyme

DNA polymerase). The primary feature that Millikan mentions to distinguish a copying process from other causal processes is that in a copying process, the individuals are "produced from one another" – that is, from other members of the same natural kind. Clearly, "copying" is being used in this context not to denote a process imbued with intentionality but one of causal replication. In this case, members of a certain chemical kind (DNA) are produced as a result of causal interactions that involve other members of that kind, rather than independently of other members of that kind. At some point in the distant past, at least one molecule of DNA must have been produced without a process of replication. But since then, DNA has always come into existence in conjunction with other members of its own kind, and this is the causal process behind most copying in the biological realm. Moreover, the primary feature of copied kinds may obtain more widely in the chemical and even physical realms. Atoms of most elements in the universe were created either by the process of primordial nucleosynthesis, soon after the big bang, or in stars, as a result of nuclear fusion and fission. Some of these reactions involve atoms of the same kind. For instance, most of the helium-4 in the universe is created by combining two helium-3 nuclei, a reaction that emits two protons (Ball 2004, 107). Thus, atoms of the most stable and prevalent isotope of the element *helium* are created as a result of a process that involves atoms of the same chemical element (of a different isotope). All atoms of heavier chemical elements are synthesized in a similar manner, either from atoms of other chemical elements or from atoms of isotopes of the same chemical element – that is, members of their own kind. The other two features of copied kinds that Millikan mentions are also satisfied in some of these cases. In the case of DNA, members of the kind are produced as a result of the same "historical environment," since all molecules of DNA descend ultimately from the same source and have a common history, and they serve the same "function," in the sense of having an effect that raises the probability that its cause will be reproduced. Even in some cases of nucleosynthesis, the effect of producing atoms of a certain element will raise the probability of producing more atoms of that same element. Moreover, the history of chemical elements is the history of the universe since the big bang, so the third feature may be trivially satisfied. Hence, the key features of copied kinds are present in some chemical and physical kinds. Second, when it comes to naturally selected biological kinds like species, though they evolve and retain the characteristics that they do partly as a result of a copying process, as Millikan points out, they also remain in (relative) homeostasis due to environmental selective pressures.

Indeed, in sexual species the whole point of the copying process is that it is not perfectly faithful and does not produce a series of identical individuals. When the copying process leads to fitness-enhancing traits, they are likely to be preserved, whereas when it leads to fitness-reducing traits, they tend to be eliminated.[10] Third, even in the social realm, there are kinds that do not seem to be etiological or arise as a result of a copying process. Just as the examples of convergent evolution cited in the previous section attest to the fact that some natural kinds of biological organs and of phenotypic traits have arisen independently in different species, at least some social kinds do not appear to arise straightforwardly as a result of a copying process. Kinds of social institutions such as *marriage* and *government* may not always arise in human societies due to a process whereby successive generations imitate the arrangements made in previous generations, but sometimes as a consequence of human propensities and dispositions. Some of these may be innate, while others may be learned, and learning is, of course, a kind of copying process. Copying is involved in some way, but it may not be a matter of directly copying these social institutions themselves from one society to another. If these institutions arise spontaneously in isolated societies, they are at best copied *within* those societies across generations but not copied from one society to another. Rather, they emerge as a result of causal processes that can be captured in the form of empirical generalizations concerning the human realm. To sum up, it seems clear that at least some physical-chemical kinds arise as a result of something akin to a copying process. Meanwhile, some biological and social kinds are not etiological or copied kinds, and even those that are etiological and partly copied are also fashioned partly by other causal processes and natural laws.

But the conclusion that biological and social kinds neither are all copied kinds nor are they the only copied kinds may seem a bit hasty. Is it that at least some biological and social kinds would not, or perhaps could not, be manifested were it not for a copying process, whereas all physical and chemical kinds at least *could* have? It is true that it is highly improbable that a tiger would arise in nature without a copying mechanism, but the same may go for some exceedingly complex synthetic chemical compounds. It may be that in the realm of biological species, there are fewer constraints on the kinds of organisms that are possible given natural laws and causal processes by comparison with constraints on kinds in the chemical and

---

[10] To agree with this, it is not necessary to believe in biological laws; there are plenty of other special-science laws that constrain the process of natural selection.

physical realms. There are presumably many more possible biological kinds than are actually manifested, but there may not be many more possible chemical elements than are actually found in the universe[11] (though there are surely many more chemical compounds). In other words, the difference between the biological-social realms and the physical-chemical realms may be that the former are less constrained than the latter in terms of the kinds that can arise there, but this claim is hard to make with precision. Anyhow, it is a different claim from the one with which we began, and it does not have any obvious implications for the presence of natural kinds in the biological and social realms.

In showing that the divide between eternal and copied kinds does not correspond neatly to that between physical-chemical and biological-social kinds, we have already encountered some evidence to suggest that the distinction is not as rigid as all that. When Millikan proposed her distinction between eternal and copied kinds, she was careful to insist that the latter do not come about *purely* as a result of a copying process. As I have emphasized, she pointed out that there were two other features or "causal factors" that typically conspire together with the copying process to maintain such kinds. The first is the environment (broadly construed) in which the copying process is carried out, which, as I already mentioned, plays a decisive role in the evolution of biological organisms and is what drives the evolutionary process forward. The second (related) causal factor is the *function* or functions of the members of the kind, in the sense of an effect that raises the probability that its cause will be reproduced. Were it not for these two complementary causal factors, members of the copied kind would probably not continue to be manifested. The reason that some things are copied while others are not has to do with causal processes that are relatively independent of the copying process itself. So other causal processes (and the laws of nature that govern them) are crucial in determining members of which kind get reproduced and perpetuated and which do not. When it comes to biological species, this is primarily the causal mechanism of natural selection. Could there be *pure* copied kinds, ones that persist solely because their members are reproduced as a result of a copying process regardless of the functions that they might have or how they interact with their surroundings? In principle, there could be kinds, members of which exist solely as a result of a copying process, but they seem to be few and far

---

[11] It is an open question whether there is an "island of stability" beyond the known chemical elements that could contain a variety of new, stable isotopes with atomic numbers much larger than those of the known elements.

between. Kinds of human artifacts, such as *1957 Eames desk chairs* or *1969 Plymouth Valiants*, may seem like plausible candidates. But even when it comes to these artifacts, what enables their members to be produced in the first place and what ensures that they continue to be copied (or not, as the case may be) is something about their functions (e.g., utilitarian and aesthetic properties) and about selective pressures in the environment (e.g., the economic marketplace). It is not as if the copying process (industrial manufacturing) would copy any old thing.[12]

There would seem to be few if any kinds that persist *purely* due to a copying process; other causal processes are always at least partly responsible for the persistence of a kind and the continued existence and production of its members. Hence, the division between eternal and copied kinds is not as sharp as it may have initially appeared. The upshot is as follows: Natural laws and causal processes generally determine what kinds of individuals can come into being in the universe. Some of these individuals arise independently of other members of their kind, while others arise partly as a result of interaction with other members of their kind, either by being reproduced by them or copied from them by a designer, or by some other process. Though the processes of reproduction and copying do play a larger role in the biological-social realm than in the physical-chemical realm, they do not operate exclusively in that realm nor do they apply to all kinds in that realm. Given that the distinction between eternal and copied kinds is not as sharp or far-reaching as it may have initially seemed, it does not provide a way of sharply distinguishing biological and social kinds from other natural kinds, nor does it give us grounds for deeming the latter not to be natural kinds.

4.5 MIND-DEPENDENT AND INTERACTIVE KINDS

One common way of distinguishing natural from nonnatural kinds is by appealing to the human-dependence of the latter. But strict independence of human beings cannot serve as a criterion for the reality of kinds, since that would make such kinds as *sickle cell anemia* and *Homo sapiens* nonreal. A more promising criterion would advert more specifically to independence of the human *mind* rather than independence of human beings more generally. Indeed, mind-independence has often

---

[12] It may be that some artifactual kinds arise purely as a result of copying, and this may make them different from many natural kinds, but I will not try to address the issues raised by artifactual kinds in this book.

been taken to be criterial for realism, whether about individuals or kinds. As Devitt (2005, 768) puts it:

The general doctrine of realism about the external world is committed not only to the existence of this world but also to its 'mind-independence': it is not made up of 'ideas' or 'sense data' and does not depend for its existence and nature on the cognitive activities and capacities of our minds.[13]

But there are seemingly bona fide mental or psychological kinds that pertain to the human mind, and hence are dependent on the human mind, which cannot be dismissed out of hand as nonreal or nonnatural – for example, *Alzheimer's disease*. It is not just that this criterion would disqualify some putative natural kinds; that may not be sufficient reason to dispense with such a hallowed conception of what is real. Rather, the more telling point is that humans and their minds are part of the universe and that there is no reason to prejudge the issue as to whether some types of human mental states, actions, and processes can be natural kinds. Instead of ruling them out as being ineligible candidates for natural kinds, it should be possible to distinguish kinds that are mind-dependent in the sense of being subjective from those that are mind-dependent merely in the sense of pertaining to the mind or being a causal product of the mind.[14]

Is there a prospect of formulating a more nuanced version of this criterion of realism? Some philosophers have been tempted to say that what matters for realism is not *causal* dependence of a kind on human minds, which would not be problematic, but *constitutive* dependence, which would be. This is what Boyd (1989, 22) appears to be getting at in the following passage:

The realist differs from the constructivist in that (like the traditional empiricist in this instance) she denies, while the constructivist affirms, that the adoption of theories, paradigms, conceptual frameworks, perspectives, etc. in some way constitutes, or contributes to the constitution of, the causal powers of, and the causal relations between, the objects scientists study in the context of those theories, frameworks, etc. The realist does not deny (indeed she must affirm) that the adoption of theories, conceptual frameworks, languages, etc. is itself a causal phenomenon and thus contributes causally to the establishment of, for example,

---

[13] This formulation is ambiguous between individuals or objects and properties or kinds, but I take Devitt to be referring to both.

[14] One possible fix would be to retain this criterion but add an exception for kinds pertaining to humans and their minds, but then we would need a realism criterion for those kinds – quite apart from the fact that such a rider would be *ad hoc*.

those causal factors which are explanatory in the history of science and of ideas. What she denies is that there is some further sort of contribution (logical, conceptual, socially constructive, or the like) which the adoption of theories, etc. makes to the establishment of causal powers and relations.

But though Boyd has drawn our attention to what seems like a promising contrast, it is not clear how to make it precise. In particular, the "further sort of contribution" that Boyd mentions is not easily separated from the causal contribution that he thinks the realist is committed to affirming. He refers to it both as "constitut[ing]" and as making a "logical, conceptual, [or] socially constructive" contribution, as opposed to a causal one, but he does not appear to explicate the nature of this constitutive contribution any further. Indeed, this alternative type of contribution is not easily ascertained. Consider again research on the subject of Alzheimer's disease. Human minds and mental activities are causally involved in the manifestation of Alzheimer's disease. To cite just one recent finding, bilingualism in some subjects has been found to delay the onset of Alzheimer's disease (Craik *et al.* 2010), which means that the activities of human beings and their mental operations play a causal role in the development of this type of disease in individual subjects. Indeed, one could imagine monolinguals (or their parents) at an early stage of their lives taking note of this finding and studying a second language in order to postpone the onset of the disease. If these individuals then succumbed to the disease later than they otherwise would have, would that make *Alzheimer's disease* not a natural kind? It seems obvious that it would not, for this is an innocent causal contribution on the part of humans and their minds, rather than a problematic one. What would be problematic? Presumably, what we want to avoid is a situation in which, roughly speaking, the researchers who investigate the kind associate whichever properties they like with the kind and that this then becomes the scientific theory about the disease and is used to determine whether someone has the disease or not. However, this is just to say that they ought to follow the evidence rather than their preconceptions, which is a sensible scientific maxim, but it is one that would seem to be applicable in all realms of human inquiry, not just the biological and social realms. Scientific investigators inquiring into the nature of kinds ought to be guided by the phenomena.

It could be objected here that it is not just a matter of due diligence on the part of investigators, for when it comes to social kinds in particular, the nature of the phenomena is such that there is a serious danger that the kinds are shaped by the beliefs of researchers. In numerous works, Hacking (1986/1999, 1988, 1991b, 1992, 1995a, 1995b, 1999) has drawn attention to such an occurrence. Hacking's version of the mind-dependence worry is that social (or human) kinds are "interactive." Though it is difficult to do justice to his concerns in a brief discussion, especially given the many detailed case studies he has elaborated in defense of his position, I will attempt a thumbnail sketch. Hacking observes that many human kinds participate in a "looping effect": Once we have identified them and described some of their properties, our words and actions precipitate a change in the phenomena being studied, which ends up altering some of the very properties that we associated with the kind in the first place. This alteration in the properties associated with the kind then "loops back," since we need to modify our initial beliefs about the kind to conform to the new reality. In the most extreme cases, the alterations may be so significant as to result in the manifestation of an entirely new kind. This process may be iterated indefinitely. One of Hacking's most encapsulated descriptions of this process at work is sketched out in the following summary of the creation of the kind *child television viewer* (1999, 27; original emphasis):

Once we have the phrase, the label, we get the notion that there is a definite kind of person, the child viewer, a species. This *kind* of person becomes reified. Some parents start to think of their children as child viewers, a special type of child (not just their kid who watches television). They start to interact, on occasion, with their children regarded not as their children but as child viewers. Since children are such self-aware creatures, they may become not only children who watch television, but, in their own self-consciousness, child viewers. They are well aware of theories about the child viewer and adapt to, react against, or reject them. Studies of the child viewer of television may have to be revised, because the objects of study, the human beings studied, have changed. That species, the Child Viewer, is not what it was, a collection of some children who watch television, but a collection that includes self-conscious child viewers.

This is clearly meant to be an illustrative and partly speculative example, rather than an in-depth historical analysis of the development of the kind *child television viewer*. For a fuller and less hypothetical account of the evolution of an interactive kind, we would have to turn to Hacking's treatments of such kinds as *multiple personality*, *fugue* ("mad traveler" syndrome), and *child abuse*. But it will do for our purposes to illustrate the main features of the process that Hacking is concerned with. (In utilizing this example, no assumption is being made about whether or not *child television viewer* is a natural kind. In fact, that is precisely what is at issue. We want to know whether the interactivity to which Hacking has drawn our attention bars such categories from corresponding to natural kinds, as Hacking himself seems to think.[15] Of course, this particular category may also fail to correspond to a natural kind for other reasons.)

In barest outline, the process might unfold as follows.[16] Researchers begin by investigating a phenomenon such as the television-viewing habits of members of a certain society. Based on their initial observations, they devise a category, with which they associate a certain extension and about which they have some beliefs (or equivalently, with which they associate certain properties). They introduce a label for this category, "child television viewer," and go on to make additional discoveries concerning the properties associated with this category (e.g., child television viewers read less than their peers, there are more overweight individuals among child television viewers than in the general population of children, and so on). As they do so and their findings are disseminated, the results of their research affect the individuals who are categorized, who then alter their behaviors in certain ways. These behavioral alterations eventually result in the people categorized having different properties. These changes are in turn duly registered by the researchers, who go on to modify the beliefs that they associate with the category. In sum, the process that the researchers set in motion when they first began to study the phenomenon led to an alteration in the phenomenon being studied that eventually necessitated revisions in the beliefs of the researchers. This looping can go on indefinitely.

In considering Hacking's treatment of the looping effect, one issue that needs to be clarified has to do with the awareness on the part of the subjects categorized. In the passage from Hacking cited above, it would seem as though the children's awareness of the label is essential for the looping effect to take place and for the kind to be interactive. But it is clear from several other passages in Hacking's writings on this topic, that awareness on the part of those categorized is not a necessary condition for the looping effect.[17] More importantly, as I have argued elsewhere (Khalidi 2010), the distinctive features and broader implications of the looping effect can be in place whether or not there is awareness on the part of those categorized. To make this more plausible, imagine that in the

---

[15] However, in more recent work, Hacking (2006) rejects the whole notion of natural kind, saying that it has done more harm than good in contemporary philosophy.

[16] The sketch in this paragraph is largely speculative, but the overall pattern is a plausible one and I take it that nothing rests on the specifics of this case.

[17] For example, Hacking (1999, 32) writes of the kind *woman refugee*: "A woman refugee may learn that she is a certain kind of person and act accordingly. Quarks do not learn that they are a certain kind of entity and act accordingly. But I do not want to overemphasize the awareness of an individual. Women refugees who do not speak English may still, as part of a group, acquire the characteristics of women refugees precisely because they are so classified."

illustration above, it is not the children themselves who are aware of the theories of the researchers but their parents. Having read reports about the research findings in the newspaper, the parents initiate certain lifestyle changes so as to alter the viewing habits of their children. They cancel their cable television subscription, acquire a public library card, arrange more frequent outings to the park, and so on, in such a way that the properties originally associated with this category of children change, and with them the beliefs of the researchers (who are conducting an ongoing study of the phenomenon). Suppose that the children have no awareness that they have been labeled in a certain way or that some researchers have theories about them. The beliefs associated with the category would still influence members of the kind and alter the properties associated with the kind (thanks to the behavior of the parents). These alterations would then "loop back" to effect a modification in the beliefs associated with the category, and so on.[18]

Hacking has identified a specific type of influence that human minds and their activities have on certain human or social kinds. On his account, their being under the influence of the theories and categories of researchers in this way precludes their being natural kinds. I would like to pursue two questions concerning interactive kinds to determine whether this phenomenon would indeed prevent social kinds from being natural kinds. First, are all human or social kinds interactive kinds, and are all interactive kinds human or social kinds? Second, what is it about interactive kinds that would imply that (all or some) human kinds are not natural kinds?

Concerning the first point, I do not think it was Hacking's intention to intimate that all human kinds are interactive kinds, but he does think that all interactive kinds are human kinds.[19] By contrast, I have argued (Khalidi 2010) that at least some nonhuman kinds can be fully interactive in Hacking's sense – for instance, biological organisms that evolve as a result of artificial selection.[20] A clear example of this phenomenon is the evolution of the domestic dog (*Canis familiaris*), which can be shown to satisfy the main conditions of the looping effect. Though a category and a label were probably not introduced when the process of domestication first

---

[18] One might still object that there has to be awareness of the category on the part of *someone* in society. But this is true of all categories, and can therefore pertain to nonsocial categories as well. For further discussion, see Khalidi (2010).

[19] For example, Hacking (1991a, 258) writes that human kinds "are different from what philosophers call natural kinds because they interact with the very beings to which they apply."

[20] This argument has precursors in several other responses to Hacking – for example, Douglas (1986), Bogen (1988), and Haslanger (1995).

began, at some stage humans began to select certain individual animals to mate with others based on the characteristics that they wanted in their offspring. They then observed the next generation of animals to determine whether those traits were in evidence and repeated the process over successive generations. Dogs are not aware of the labels that human breeders attach to them or the beliefs that they have about various breeds, the desirability of certain traits, and the alterations they intend to make in the dog lineage. But I have already mentioned that awareness on the part of the individuals classified is not necessary for the looping effect to get underway. It may be conceded here that interactive kinds are not just human kinds but that they also include some nonhuman biological kinds (and perhaps others). Indeed, it may be that all interactive kinds are historical (or copied) kinds in Millikan's sense (see section 4.4), though the converse is not true. Nevertheless, it might be maintained that interactive kinds, whether human or not, are not capable of being natural kinds.

What is it about interactive kinds that would disqualify them from being natural kinds? Consider more closely the process of artificial selection of animals like dogs, in which human beings carefully choose individuals with certain desirable traits to mate with other individuals with the same or other desirable traits. These phenotypic traits may be morphological (e.g., color, size, ear shape), behavioral (e.g., obedience, docility), or functional (e.g., hunting abilities). The offspring are then observed and, depending on their traits, breeders will again select some of them to mate with others, and they will once again ascertain which traits are manifested in the next generation. Based on a (possibly long) sequence of iterations, they will effectively have fashioned the individuals that result, and these individuals may well come to belong to a different kind from the one they started with (as *Canis familiaris* diverged eventually from *Canis lupus*, though opinions differ as to whether it is a distinct species or merely a subspecies). In this process, human beings are involved at every stage in selecting the individuals that have the phenotypic traits that they value, and they mold and sculpt future generations of animals to suit their needs. Even though the end product will have been carefully shaped by an iterative process of human intervention that responds during each iteration to the properties that result in the subsequent generation of organisms, that should not disqualify it from being a genuine natural kind. After all, artificial selection is at bottom a causal process much like natural selection, and the fact that human design is involved should not prevent it from giving rise to natural kinds. Breeders may start out at the beginning of this process by saying that their aim is to produce a new kind of animal with such-and-such properties, and they may indeed end up with the very same properties that they intended to produce. In one sense, the kind *dog* (*Canis familiaris*) is the result of human action and belief (e.g., as to what the desirable traits are in a domestic animal). There are dogs in the world at least partly because humans wanted there to be members of a kind with such-and-such properties. But it is not that thinking that something is a dog makes it a dog, or that any individual organism is a dog merely because humans (collectively or individually) think it is one. Again, the mind-dependence, while very real and instrumental, does not seem to be of the pernicious variety that would pose a threat to realism. Moreover, although the looping effect is sometimes described by Hacking as playing a consti-tutive or conceptual role in influencing the kind, Cooper (2004) has argued plausibly that the phenomenon can be captured in causal terms. The looping effect would seem to be a real causal pattern, though it violates the mind-independence criterion for realism since an interactive kind does indeed "depend for its existence and nature on the cognitive activities and capacities of our minds" (Devitt 2005, 768).

To see further that the mind-dependence of some kinds or their manipulability by humans does not necessarily impugn their reality or objectivity, consider another case in which human beings play a vital role in the molding and sculpting of a kind of chemical compound, the process referred to as "*in vitro* selection." In this process, scientists select fragments of nucleic acids (single-stranded DNA or RNA) that bind to specific organic compounds (target molecules). The process is an iterative one, in which scientists gradually select the nucleotide with the highest binding affinity to the "target" molecule. They start with a very large "library" of all possible sequences and after, say, ten iterations end up with a far smaller set of nucleotides that will serve the purpose that they had originally intended – for example, a pharmaceutical product that will fulfill a highly specialized aim. Here, we have manufactured a new kind of chemical compound according to our specifications, by means of an iterative process in which each iteration modifies the compound based on the properties that it already has. The end product will presumably have many of the properties that were identified at the beginning of the process when the target molecule was specified, along with others that were not, leading the researchers to modify their beliefs, and perhaps driving them to make further interventions on the compound in order to modify it further to suit their purposes.

What we are left with is a phenomenon, the looping effect, that pertains primarily to the social and biological realms (though it can also arise in other realms, as I have just argued) in which deliberate human intervention plays a significant causal role in determining the properties associated with a certain kind of individual, sometimes culminating in the creation of an entirely new kind of individual. Kinds that are subject to the looping effect are under human control and are therefore causally mind-dependent, but they certainly do not exist solely in the mind of the beholder, and the individuals that belong to those kinds do not do so simply as a result of human decision. The kind itself and membership in the kind are objective matters that are not mind-dependent in the problematic sense of being subjective.[21] (This conclusion will be further justified in section 6.5.)

4.6 INSTITUTIONAL AND CONVENTIONAL KINDS

In the previous section, I argued that although some social and biological kinds (and perhaps other kinds) are in some sense mind-dependent and interact with our beliefs about them, the type of mind-dependence that they exhibit does not render them subjective and does not undermine a realist attitude towards them. But I have not tried to further articulate the more pernicious sense of mind-dependence or to spell out in more detail what it would be for kinds to be subjective. In an influential discussion, Searle (1995) has tried to do just that. He argues that although social kinds are not epistemically subjective, they are ontologically subjective. But I will try to show that the problem with Searle's position is that it concedes too much to the antirealist about social kinds and, moreover, the account that he gives of social kinds would appear not to apply to many such kinds. I will also argue that the social kinds to which Searle's account applies most aptly may not be natural kinds for other reasons than the ones that Searle suggests.

In his discussion of social kinds, Searle (1995, 1) begins with a puzzle: How can there be "things that exist only because we believe them to exist" (e.g., money, property, governments, and marriages), yet many facts about these things are objective facts? To resolve the puzzle, he distinguishes two kinds of subjectivity, ontological and epistemic subjectivity. Epistemic subjectivity applies primarily to *facts* and *judgments*: A judgment is

---

[21] There is another problem associated with interactive kinds that pertains to their epistemological rather than their ontological status. Hacking (2006) has called them "moving targets," drawing attention to the difficulties associated with studying a kind whose properties are continually changing. While this may complicate the efforts of scientists to study such kinds, it does not directly affect the possibility of their being natural kinds. Also, see Mallon (2003) for an argument that the looping effect does not completely destabilize kinds and prevent scientists from studying them; indeed, in some instances it serves to stabilize the kind.

epistemically subjective when its truth or falsity depends on certain mental attitudes or feelings, whereas it is epistemically objective when it does not exhibit such dependence. By contrast, ontological subjectivity applies to *entities* and *types of entities*: Subjective entities (tokens or types) are ones whose mode of existence depends on being felt by subjects (though he might have added, thought or otherwise mentally apprehended), whereas objective entities do not exhibit such mental dependence. The resolution of the puzzle lies in seeing that although social kinds (like *money, marriage*, and *elections*) are ontologically subjective, judgments about them can be epistemically objective. Thus, the kind *money* is ontologically subjective, since its existence depends on our mental attitudes, but the judgment that this coin is an instance of money, or that that is a ten-dollar bill is epistemically objective, since the truth of these judgments is independent of our attitudes.[22]

While I agree with Searle that facts and judgments about social kinds can be perfectly epistemically objective, I am not ready to give up so easily on the ontological objectivity of social (and other mind-dependent) kinds. Moreover, it is not clear from Searle's account how one can separate the two types of objectivity, for it is not obvious that the very existence of something can depend on mental attitudes although facts about it do not. I have already argued in the previous section that at least one kind of mind-dependence of many social (and other) kinds is innocuous and ought not to preclude a realist attitude towards them. In Chapter 6, I will try to propose an understanding of the objective–subjective dichotomy that leaves room for the objective existence of social kinds. But before doing so, I will argue in this section that Searle's overly narrow view of social kinds is what has led him to abandon ontological objectivity for social kinds. This will also help to elucidate the nature of social kinds and the particular type of mind-dependence that they exhibit.

For Searle, the paradigmatic examples of social kinds are those that have an overtly or covertly conventional or institutional nature, such as *money, marriage, private property*, and *elections*. Such social kinds have the features that they do as a result of conventions, rules, or laws that outline the conditions that need to be fulfilled for something to count as a member of

---

[22] One might wonder: *Who* needs to have the appropriate mental attitudes to money for it to be money (e.g., the majority of a society, the experts in that society, etc.)? On Searle's account, these social kinds depend for their existence on "collective intentionality" and he has a peculiar view of what collective intentionality amounts to. However, I will bracket this issue since I do not think it matters for our purposes, as long as these social kinds are in some way dependent on mental attitudes, whether individual or collective.

that kind. Searle (1995, 32) says that what it is for some social kind, *x*, to be *x* is simply to be regarded as, used as, and believed to be *x*. For example, what it is for something to be a ten-dollar bill is for it to be regarded as such, to be used as such, and to be believed to be such, and what it is for something to be a cocktail party is for it to be believed to be such, considered as such, treated as such, and so on. He observes that this holds for the type (or kind) as a whole, though not always for each individual token of the type. In the case of the *ten-dollar bill*, the type as a whole needs to be regarded as such, though there may be an individual token that is never put in circulation and is never actually believed to be a ten-dollar bill by anyone (suppose it drops straight from the printing press into a crack in the floorboards). Nevertheless, it would still be a ten-dollar bill. But in the case of the *cocktail party*, each individual token needs to be regarded as such – otherwise it would not be a cocktail party.

Searle has identified an interesting facet of some social kinds and much of his analysis applies well to the kinds that he discusses. But if his account is meant to be a general account of social kinds, it seems not to describe many if not most social kinds. Institutional or conventional kinds tend to have the character that Searle describes, but other social kinds that are not as rule-governed or convention-bound do not conform to the same analysis. This point has been made convincingly by Thomasson (2003a, 2003b), who criticizes Searle for failing to recognize that many social kinds do not depend for their existence on people's having thoughts about those kinds themselves. As she points out, this may hold true of the kind *money*, but not of the kind *recession* (cf. *inflation, racism, poverty*). Thomasson (2003a, 276; original emphasis) writes that "a given economic state can be a recession, even if no one thinks it is, and even if no one regards *anything* as a recession or any conditions as sufficient for counting as a recession." Unlike money or cocktail parties, neither the type as a whole nor any particular token need be regarded as such to count as a recession.

Institutional or conventional kinds seem to be dependent on human beliefs and attitudes more directly than other social kinds. They are dependent not just on the existence of human beings, or the existence of human mental states generally, but also on the existence of specific mental states that pertain to those kinds themselves. In a sense, they would seem to be even more mind-dependent than Hacking's interactive kinds; not only can they interact with human mental attitudes, but they also require certain very specific attitudes to be in place for their very existence. (Note that institutional kinds may be interactive though they need not be, and interactive kinds may be institutional though they need not be.) However,

I want to argue first that the degree of dependence on specific human mental states may be somewhat exaggerated even when it comes to some of these institutional or conventional kinds, and second that mind-dependence is not what prevents some of these kinds from being natural kinds.

To support his view of social kinds, Searle (1995, 33–34) describes a cocktail party gone wild:

> If, for example, we give a big cocktail party, and invite everyone in Paris, and if things get out of hand, and it turns out that the casualty rate is greater than the Battle of Austerlitz – all the same, it is not a war; it is just one amazing cocktail party; part of being a war is being thought to be a war.

Here, Searle seems to be insisting that, at least for some social kinds, individual instances of that kind need to be regarded as such by some (or perhaps most) people to be members of that kind. But we have also seen that Searle allows that some individual instances of a social kind need not be regarded as instances of that type to qualify as instances of the type (the case of the ten-dollar bill that drops through the floorboards). Why then does he not allow it in this case? Elsewhere, he states that it is possible for a token not to be considered a member of a type when that type is "codified in an 'official' form" (1995, 53). This would not appear to hold of cocktail parties and wars (outside certain restricted contexts); for these social kinds, each token needs to be thought of as being an instance of that type, in addition to the type itself being regarded as such by members of society. It is true that in many cases, part of what it is to be a war is being thought to be a war and part of what it is to be a cocktail party is being thought to be a cocktail party. However, it does not follow that each individual war (cocktail party) is such so long as everyone considers it to be a war (cocktail party), nor that no individual event could be a war (cocktail party) unless somebody considered it to be a war (cocktail party). In the case of Searle's Parisian bash, it may be reasonable to say that it was *intended* as a cocktail party but it ended up as a street fight or a brawl (though perhaps not a war). Even if the organizers and participants all continue to insist that it was just one big chaotic cocktail party, it would not be absurd for an observer (perhaps a later historian or sociologist) to conclude that it was really a street fight or a brawl. Similarly, we might say in retrospect that the border skirmish between Ruritania and Lusitania was really a short war, though it was not considered as such by anyone at the time (and even our saying it in retrospect is not necessary for it to be a war). At least for some social kinds of the conventional or institutional variety, even if all social actors agree that something counts as a token of social kind *K*, that does not seem to guarantee that it is indeed a member of kind K. And even if no one regards something as a token of social kind K, it may well be a member of kind K. Note that this is different from Thomasson's point, which was discussed above. She argues correctly that when it comes to many if not most social kinds (e.g., *recession*) we need have no attitudes towards them for them to be the kinds that they are. My point is that even for some of Searle's institutional or conventional kinds, for which some attitudes need to be in place concerning the type, these attitudes may not need to be in place for each token of the type: We may all have the attitude that it is a cocktail party when it is not really a cocktail party, and we may all have the attitude that it is not a war when it really is a war. But the larger point is not that there are exceptions but that this shows that these kinds are not purely conventional, but at least partly *causal* in nature. I would venture that we judge whether or not the kind has been manifested at least in part by what kinds of causes and effects it has.[23]

Searle's thesis holds more nearly when it comes to social kinds that are purely of a conventional nature; that is, kinds whose associated properties or conditions of membership are more strictly laid out in a set or rules or laws. Consider a category like *permanent resident*; in many jurisdictions, require-ments are set out that specify what conditions one has to satisfy to be a permanent resident of that jurisdiction (e.g., that one not have a criminal record). To be sure, it is not the attitudes of members of society that determine one's status as a permanent resident, for this is usually determined by the attitudes of officials of the state, who are informed by the appropriate laws and statutes. Officials may make mistakes concerning the conditions that a person meets (e.g., they may think that she has a criminal record when she does not), but even when they do so, it is usually their beliefs or say-so that determines whether someone is a permanent resident or not. In such cases, it is the beliefs of officials – derived from an explicit convention or law – that determines whether someone is or is not a permanent resident.

Based on this discussion, I would conjecture that we can plot social kinds on a rough continuum in terms of their degree of conventionality,

---

[23] Note that it would be an exaggeration to say that there are no physical constraints on the tokens of a conventional or social kind. *Money* would seem to lend itself well to this claim, but even money cannot very well be made out of ice (at least not where temperatures often rise above 0 °C), or a radioactive isotope with a very short half-life, or a rock the size of the moon. So the nature of these kinds is somewhat constrained by factors other than the attitudes of human beings. An argument along these lines has also recently been made by Guala (2010, 260), who writes with reference to the kind *money*: "What counts as money does not depend merely on the collective acceptance of some things as money, but on the causal properties of whatever entities perform money-like functions."

from those whose nature is not conventional at all, in the sense that human beings need not have any attitudes towards them for them to be manifested (e.g., *recession, inflation*), to those whose nature is somewhat conventional, in that it is at least partly determined by specific attitudes that human beings have towards them, though these attitudes are neither strictly necessary nor sufficient for their manifestation (e.g., *war, cocktail party*), to those whose nature is more purely conventional, in the sense that their existence is almost entirely determined by specific attitudes that human beings have towards them (e.g., *permanent resident*). But even when it comes to these last kinds, which are most conventional, though they are more directly mind-dependent (and are so in a different way from Hacking's interactive kinds), it is not their mind-dependence as such that may rule out a realist attitude towards them. I will now try to explain and justify this claim.

The first category of social kinds is mind-dependent in the sense that at least some human mental attitudes need to be in place for the kind to exist at all and for specific instances of the kind to exist, but they need not be directed towards the kind itself. There could not be such a thing as a *recession* unless human beings bought and sold commodities and had attitudes towards certain goods and services, but they need not have any attitudes towards recessions. Also, a specific period in the life of an economic market could be considered a recession only if human beings were engaged in economic activity during that period, and this in turn requires certain mental attitudes. The second category of social kinds is mind-dependent in a stronger sense; here, certain specific attitudes towards the kind itself need to be in place for the kind to exist in the first place, but individual members of that kind might come into being without those attitudes being manifested towards them. This would be the case for *money* and perhaps also for *wars* and *cocktail parties*. For wars to exist at all, there need to be attitudes involving the concept *war* and related categories. But it may be that any individual war could be a war without its being considered as such by the parties (or indeed anyone else). In the third category are social kinds whose existence depends on specific attitudes towards the kind involved, and whose individual instances must also be deemed by at least some people to be members of the kind for them to be members of the kind. To have a category of *permanent resident* in a certain state requires there to be conditions set out by officials of that state and attitudes involving the category *permanent resident*. Moreover, no individual could be a permanent resident of that state without certain officials having the requisite attitudes towards *them* (namely, that they satisfy the conditions).

What I have been calling purely "institutional kinds" or "conventional kinds" coincide with the third category of social kinds, which are mind-dependent on two counts: (a) The mental attitudes upon which they depend involve the kind itself, and (b) both types and tokens depend on mental attitudes. The first category of social kinds does not satisfy either condition (a) or (b), the second category satisfies (a) but not (b), and the third category satisfies both (a) and (b). The mind-dependence of the third category of kinds is of a different type from that pointed out by Hacking,[24] but here again, the mind-dependence does not seem to undermine the reality of such social kinds. Rather than their very mind-dependence or the nature of their mind-dependence, I would propose a different reason as to why social kinds in the third category, at the most conventional end of the scale, are not likely to be natural kinds. I have been arguing that real properties and natural kinds participate in real causal patterns, which are projectible and can be captured in causal laws and generalizations. There is no reason to think that such patterns do not exist in the social or human realm as in the nonhuman realm, and some of these causal patterns involve human beings and their minds. But when it comes to the most conventional social kinds, the properties associated with them are so associated because they are explicitly codified in a set of rules or laws. Therefore, if a conventional kind K is associated with a set of properties, that is not because there are *causal* connections between these properties. Rather, they are associated with one another because a social institution or community has decided to associate these properties with the kind. Any relations between these properties are not the outcome of causal processes but are the result of a conventional linkage between them. For example, the property of not having a criminal record is associated with the kind *permanent resident* by law or statute, not as a result of causal linkage. That is why the most conventional of social kinds are generally not natural kinds (I will try to justify this further in section 6.3). Still, it is not that these kinds *cannot* participate in causal processes; in fact, there are perhaps two principal ways in which they do so. First, it may be that their associated properties and conditions of membership have been codified by causal relations and patterns that existed before the rules and regulations were

---

[24] Hacking's interactive kinds would seem to be capable of fitting into any of the three categories. However, they appear to be distinguished from other mind-dependent kinds in that: (i) The mental attitudes involved are theoretical or classificatory, and (ii) the mental attitudes modify instances of the kind (and therefore the kind itself) over time. But then perhaps they are less likely to be drawn from the first category, since the mental attitudes are classificatory and therefore presumably concern the kind itself.

drawn up (this may be the case for categories such as *war*, though I have argued that it is not at the most conventional end of the scale). Second, once the conditions for membership in the kind have been fixed, the properties associated with the kind may come to participate in new causal patterns that were not in existence before the creation of the conventional kind. For instance, we might discover that most permanent residents of a certain state are urban dwellers; this would be quite different from "discovering" that most permanent residents do not have a criminal record at the time of becoming permanent residents. In this case, even though the kind originated as a conventional kind explicitly associated with a set of properties (e.g., lacking a criminal record), it may come to participate in certain causal processes in a particular social setting (e.g., dwelling in an urban center). The links to further properties, which were not listed among the conditions that were written into the kind in the first place, represent causal relationships that render the kind projectible in novel ways. In these latter cases, the kind may well be (or may well become) a natural kind, not because of its conventional properties but because of new causal patterns that arise, associating it with new causal properties.

Searle's account of social kinds is more applicable to those social kinds that can be placed towards the conventional end of the scale. If there are such purely conventional kinds, then they may not be natural kinds, not because of their mind-dependence but because the generalizations that one can make about them are the result of rules and regulations rather than causal processes that they enter into. Their "projectibility" is a matter of convention rather than a result of causation (whether mental or nonmental). Even these purely conventional kinds, however, sometimes become implicated in causal processes. As for other, less conventional, social kinds, they may also participate in real causal processes, and to the extent that they do, their mind-dependence does not pose an obstacle to their being natural kinds.

4.7 NORMATIVE KINDS OR EVALUATIVE KINDS

There is an additional challenge that needs to be confronted in connection with the claim that human or social kinds (and perhaps also some biological kinds) can be natural kinds. Some philosophers allege that the value-ladenness or normative aspect of human kinds prevents them from being genuine natural kinds. The question of value-ladenness in the sciences, especially in the social sciences, is a vast one that I cannot treat adequately here. But I will try to address it in a limited fashion with reference to social (and perhaps other) kinds, taking my cue from an influential treatment by Griffiths (2004), who refers primarily to the category *emotion* in the social and psychological sciences. According to Griffiths, *emotion* is not a natural kind, at least partly because it has a normative or evaluative dimension. Specifically, this category is involved in moral judgments, as are other categories in the human sciences, including the social and psychological sciences but perhaps also such sciences as medicine.[25]

Griffiths' understanding of natural kinds agrees in important respects with the one I have been developing in this book, but there are a few significant differences that need to be mentioned before explicating his stance on the normative dimension of some social kinds. After stating that Mill and Whewell are the sources for his conception of natural kinds, Griffiths (2004, 903) explains his usage as follows:

I use the term 'natural kind' to denote categories which admit of reliable extrapolation from samples of the category to the category as a whole. In other words, natural kinds are categories about which we can make inductive scientific discoveries.

I am in broad agreement with this conception of natural kinds, although it has become clear in Chapter 3 and in this chapter that the projectibility of the properties associated with natural kinds ought to stem from their involvement in genuine causal processes and their participation in real causal patterns (rather than, say, being a matter of convention, as I argued in the previous section, for example). Griffiths also endorses Mill's requirement that there be an "inexhaustible" number of things discoverable about natural kinds, but as I explained in Chapter 2, that condition is too stringent and indeed perhaps unsatisfiable. On a less substantive note, Griffiths prefers not to use the term "natural kind" on account of the "historical baggage" that is associated with it, and proposes instead to use "investigative kind" or "epistemic object". But I have already explained in Chapter 2 why I prefer to reclaim the term "natural kind" rather than let it be inextricably associated with untenable essentialist claims.

Apart from these differences, the position on natural kinds that Griffiths endorses is compatible with mine, and he writes explicitly that he wants his account to serve for the biological and social sciences.[26] Moreover,

---

[25] Of course, it is possible to have moral attitudes to nonhuman animals and other entities outside the human sciences; in that case, this argument might apply beyond the human realm. But the issue raised by Griffiths pertains primarily to categories in the human sciences.

[26] For instance, Griffiths (2004, 905) states: "The traditional requirement that natural kinds be the subjects of spatiotemporally universal and exceptionless laws of nature would leave few natural kinds in the biological and social sciences, where generalizations are often exception-ridden or only locally

Griffiths (2004, 906) mentions that the interactive dimension of social kinds elaborated by Hacking (and discussed in section 4.4) does not make them unsuitable for performing the inductive and explanatory roles associated with natural kinds, though he does not argue this point at any length. However, what does jeopardize the natural kind status of some psychological and social kinds, according to him, is the fact that at least some of them have a normative dimension that undermines their inductive and explanatory role.

Griffiths' argument, which makes primary reference to the kind *emotion* as it is used in the psychological sciences (including such sciences as psychiatry and neuroscience), proceeds in three steps. First, he explains that "investigative kinds" (which I would consider natural kinds) are "open-ended" in the sense that their extension and intension can both be revised to preserve their projectibility and explanatory power. What Griffiths is referring to here is the idea that we revise our scientific categories to conform to new empirical evidence in order to preserve projectibility. This is an idea that I illustrated briefly in section 2.2 with reference to the kind *vertebrate*. As we discover more about evolutionary history and the nature of biological organisms, we revise the belief that all vertebrates have a spinal column, thereby also revising the extension of the category *vertebrate*. Griffiths adds that normative categories also exhibit this kind of open-endedness albeit for different reasons. In the case of normative categories, we revise them not (just) to preserve their projectibility or inductive power but also, according to him, "as part of a project of social reform." Griffiths (2004, 908) illustrates this by citing the case of the category of *child abuse* that has been studied by Hacking (1992; see also 1988, 1991a):

The change from viewing a pattern of childcare as normal to viewing it as abusive need not reflect an epistemic project, such as maximizing the predictive power of child abuse as a diagnostic category in psychiatry. The change can equally well represent the spread of a new normative model of the relationship between parent and child, or a change in the relative value placed on various traits of the older child or adult, such as placing a higher value on personal fulfillment and a lower value on social conformity.

The idea, in brief, is that administering corporal punishment to children came to be included in the category of *child abuse*, along with inflicting

---

valid. Fortunately, it is easy to generalize the concept of a law of nature to the notion that statements are to varying degrees 'lawlike' (have counterfactual force). This allows a broader definition of a natural kind."

serious bodily damage and also some forms of verbal abuse and neglect, because this range of behaviors came to be seen as immoral or out of keeping with social values such as personal fulfillment. The primary impetus behind these revisions was not epistemic, according to him, but moral or evaluative.

The second step in Griffiths' argument consists in noting that at least some social and psychological categories are *both* investigative kinds and normative kinds. That is, they play an epistemic role as well as an evaluative one, and these two roles cannot generally be cleanly separated, since the same categories are used to explain human behavior as well as to prescribe and condemn it. In particular, focusing on the category of *emotion* as well as some of its subcategories like *anger*, he argues that there are normative standards for when anger is warranted, in addition to explanatory theories as to the causes and effects of anger.

The third step in the argument notes that the dual role that categories like *emotion* and *anger* play makes them answerable to these different pressures, and that these pressures can push the category in different directions. Since normative judgments about emotion are not likely to be independent of our beliefs about the nature of emotion, the categories are inextricably bound up with both kinds of considerations, epistemic and moral. Though Griffiths does not say so in so many words, the implication is that the moral dimension impinges on the epistemic in such a way as to influence the shape of these categories; that is to say, their extensions and the beliefs associated with them (chiefly concerning the properties with which they are associated). This would seem to distort the categories and prevent them from accurately reflecting the relevant kinds.

If Griffiths is right about at least some categories that pertain to human beings and can have a moral dimension, then this would appear to pose a problem for my account, at least if such categories are to reflect natural kinds. I stated in Chapter 2 that categories that are projectible and feature in scientific explanations and predictions are those that correspond to natural kinds. Our best guides to nature's divisions are those categories that enable us to explain and predict natural occurrences by tracking causal patterns. Hence, categories that serve this epistemic purpose denote natural kinds. But what happens when categories that serve an epistemic purpose also serve some other purpose, say, a moral one, and when these purposes do not always coincide? Can such categories still be trusted to reflect real causal patterns?

If there are indeed categories that are answerable to pressures other than epistemic ones, then they could not be guaranteed to pick out natural kinds. In order to see whether this is a likely eventuality in the social sciences, I will try to make Griffiths' case more plausible by looking at Hacking's example of *child abuse* in a little more detail. Since I am attempting to make a case on Griffiths' behalf, I will build on some of Hacking's findings but I will also make some empirical assumptions to present Griffiths' view as positively as possible. If this were a category in the social sciences that served an epistemic purpose, we would expect it to be projectible, to enter into empirical generalizations (albeit not exception-less laws), and to supply us with explanations of social phenomena. Let us suppose for the sake of argument that this is indeed the case for the category *child abuse* and closely related categories like *child abuser* and *abused child*. But if Griffiths is right, in addition to this epistemic purpose, this category serves an evaluative purpose as well, perhaps playing the role of censuring certain behaviors. Indeed, even though Griffiths does not cite them, some of Hacking's findings about the evolution of this category since the early 1960s can be used to back up Griffiths' conjecture. Among many other observations that he makes, Hacking (1991a, 263) notes that in some legal jurisdictions in the USA, "fetal abuse" has come to be classified as a form of child abuse. Mothers who drink heavily or take certain drugs while pregnant are considered by some to have engaged in a form of child abuse. Hacking writes that this initiative has come from law enforcement officers and others who are concerned to prosecute mothers who engage in these behaviors. This implies that the drive to revise the extension of the category to include this type of behavior has originated among those wanting to impose a legal censure upon it (rather than those who are more concerned to understand and explain the phenomenon of child abuse). In this case, the evaluative dimension is legal rather than moral, but the two modes of evaluation are closely related, and Hacking implies that this legal move derives from certain social mores. "The cases that have attracted most attention have to do with crack cocaine and fit well into crack-as-a-social-problem," Hacking (1991a, 263) writes, and he also indicates his evaluative disagreement by adding: "The moves that are being made here are quite disreputable." Although Hacking does not say so, let us imagine that this value-driven extension of the category does not increase the inductive or explanatory power of the category. In other words, let us grant Griffiths the claim that revising the extension in this way is at odds with the epistemic role that the category plays and that it does not enable us to make new generalizations about child abuse or explain its associations with certain social conditions or other social behaviors. We can even suppose that it undermines the epistemic role that the category already plays, in that it prevents us from making some of the generalizations that were formerly made about the phenomenon of child abuse, or that it introduces exceptions to some of the existing generalizations. For instance, it may be that "fetal abuse" does not share many of the causes and effects that other kinds of child abuse tend to possess. If all this were granted, then it would seem as though we have a fitting instance of the possibility to which Griffiths is drawing our attention.

Griffiths' speculation about social categories is quite plausible. Whether or not the details are true of the category *child abuse*, there are grounds for thinking that at least some social and psychological categories play both an epistemic and evaluative role. There are instances in which the evaluative role of a social or psychological category is explicitly acknowledged by the researchers deploying the category. To take a particularly blatant example, in their analysis of the psychiatric category of *mild cognitive impairment* (MCI), Graham and Ritchie (2006) seem to make an argument for privileging ethical considerations over epistemic ones in delimiting this category. After acknowledging that "MCI, like the dementias that it is presumed to precede, is a heterogeneous condition with no certain biomarker or known etiology" (2006, 36), they go on to argue that the category "has played an important role, responding rapidly to changing social attitudes toward aging and a demand for increasing intervention to improve quality of life in the elderly, in the face of a lagging public health policy that has, for too long, neglected psychogeriatrics" (39). These researchers appear to admit that the category MCI does not play an epistemic role but rather an evaluative or moral one, which has to do with ameliorating living conditions for the elderly. In this case as well as in other cases, the evaluative role may well suggest a different demarcation from the epistemic one. Revisions to a category that are meant to serve one purpose can make it less well suited to the other purpose. It is not that these purposes are necessarily in conflict but that they serve different needs and can therefore come to be opposed to one another. Hence, there is a legitimate reason for thinking that some social categories do not track natural kinds when they are being put (at least partly) to evaluative uses. These uses are not always as explicit as in the above example, and epistemic and normative purposes cannot always be easily disentangled. We can grant all this, yet not give up entirely on extricating the epistemic dimension of categories from any evaluative aspect that they might have. Social scientists who are interested in understanding and explaining a phenomenon have a stake in being able to track its causes and effects. The need to delineate social categories in such a way as to identify phenomena that are genuinely causally linked is central to social science. It is not a coincidence that the move that

Hacking describes with respect to *child abuse* is introduced by law enforcement officials rather than sociologists or public health specialists. Since the purposes of the latter are primarily to understand and explain social reality, rather than to pass moral or legal judgment, any revision of the category that weakens its epistemic efficacy is liable to be resisted by at least some social scientists. Thus, if a value-driven revision of a category pulls in a different direction from the epistemic, these revisions should be visible to inquirers who are concerned to preserve its epistemic role. That is not to say that such changes will always be unmasked and rejected by the social science community. Indeed, in some cases, they may be advocating certain moral or social causes and may deploy a category even though it does not serve epistemic purposes, as in the case of MCI. But epistemic purposes are extricable in principle from evaluative ones through considerations of projectibility, explanatory efficacy, and the like. The challenge posed by evaluative categories (or the evaluative dimension of certain categories) in the social sciences is a real one; in my view, it poses the largest obstacle to the discovery of natural kinds in the social sciences. But I would suggest that it can be met by social scientists so long as they aspire to devise categories that enable them to accomplish epistemic purposes, thereby averting attempts to expand or otherwise modify categories in such a way as to serve moral interests. In fact, this seems to be what actually transpired with the attempt to expand the category of *child abuse* to include "fetal abuse", since the modification did not catch on, and "fetal abuse" is not generally considered a form of child abuse among social scientists.

So far I have assumed that social scientists are concerned to ascertain causes and effects in their domains of interest and that this would give them an incentive to disentangle the epistemic dimension of social categories from the evaluative dimension. Whether or not they will always succeed in doing so is another matter, but my contention is that the task is not a futile one. However, the claim is sometimes made that a value-free social science is not just unattainable but also undesirable. Among others, Fay (1983) has argued that the aim of social science ought to be moral as well as epistemic, specifically that the social sciences should strive to emancipate and empower human beings rather than just study them. He acknowledges that one of the primary aims of social scientists is to articulate "causal generalizations" concerning social phenomena (though not strict laws since these generalizations will be limited to a specific cultural context). But he also argues that social scientists should aim to be *critical theorists* who aim to transform society. Fay (1983, 108) explains the role of such social scientists as follows:

[A] critical social scientist actually desires to see his causal generalizations made otiose by a group of actors who, having learned them, alter the way they live. He desires this because it means that he has been successful as a theorist in helping to alter the social world which he is studying.

In this passage, Fay is drawing on a conception of social science that derives at least partly from the ideas of the Frankfurt school, about which much has been written. Without trying to do justice to this rich tradition, these recommendations for social science would seem to presuppose that one *can* engage in a type of social science that does *not* aim to rid society of domination and oppression. By prescribing a social theory that aims at a more just social order, Fay implicitly suggests that one can do social science without attempting to emancipate human beings. In fact, it would seem as though an honest effort to change oppressive social conditions would be based first on a thorough understanding of these social conditions, and this in turn means being able to articulate the causal processes that operate within them. After all, Fay (1983, 109) acknowledges that the conception of social theory that he advocates "sees humans as natural creatures in a natural world of cause and effect, and thus as fit subjects for science."[27] Thus, it would seem as though even some of those philosophers who argue that social science ought to have a moral aim or purpose can agree that it is possible to disentangle that purpose from the epistemic one of understanding the causal nature of social phenomena. If so, then it is imperative that an emancipatory project of this kind be informed by an objective account of social reality whose social categories accurately reflect causal relationships and social causes and effects.[28]

4.8 CONCLUSION

In this chapter I have discussed the main challenges that might be perceived to prevent biological and social kinds from being natural kinds. I have argued that these kinds ought to be regarded as genuine instances of

---

[27] Fay goes on to say that humans are also capable of a kind of initiative that other animals are not capable of. His point here seems to overlap considerably with Hacking's that there are interactions between humans and the categories that are used to classify them; this was discussed in the previous section, so I will not pursue it further here.

[28] Miller (2000) thinks that social kinds reflect moral interests, on the grounds that we often judge a social kind to be real because we think that individuals would enhance their self-expression were they to identify with that social kind. However, he also acknowledges that in many cases, the reality of social kinds is of a piece with the reality of nonsocial natural kinds, on the basis of their causal properties and their indispensability to adequate explanation of the phenomena. Similarly, Haslanger (2006) draws a distinction between the "operative concept" of a social kind and the "target concept" of a social kind, which I think is compatible with this position.

special-science kinds, as characterized in Chapter 3. Some of them are etiological, copied, interactive, and mind-dependent, without thereby jeopardizing their claim to be natural kinds. Although I have argued that the broad challenges represented by each of these types of kind can be met, I have not tried to show in detail that specific biological and social kinds are indeed natural kinds. I have merely tried to overcome what I take to be the main principled objections to there being natural kinds in the biological and social sciences. In the following chapter, I will look at particular case studies to show how some typical categories in these sciences do in fact correspond to natural kinds, at least in the sense that I have been defending in this book.

In arguing for the possibility of natural kinds in the biological and social sciences, one issue that has arisen in more than one guise has to do with the mind-dependence of such kinds. Some philosophers have tried to reconcile the mind-dependence of human kinds with their reality or objectivity, by distinguishing different kinds of mind-dependence (e.g., causal as opposed to constitutive), and by saying that human kinds have one kind of mind-dependence but not the other. However, none appear to have made this distinction in a rigorous way. By contrast, my suggestion would be to give up on reformulating the thesis of mind-independence, which some philosophers have regarded as being criterial for realism about entities and kinds, in such a way as to allow for the existence of (at least some) human kinds. If we want to be assured of the reality of our categories and be confident that they correspond to natural kinds, then there is no need to invoke mind-independence. It is true that our mental preconceptions may sometimes lead us astray in ascertaining the structure of reality, but so can other things too. Broadly speaking, the aim should not be to guarantee mind-independence but to ensure world-dependence. And since the features of the world that we want to capture are causal processes and relationships (including those that involve human minds), I will argue in Chapter 6 that an alternative formulation of realism towards kinds should advert to their causal efficacy rather than their mind-independence. Before doing so, in Chapter 5, I will use specific case studies drawn from the biological and social sciences (among others) to demonstrate in more detail that there are natural kinds in these realms and that they are implicated in causal processes. At least some of the categories of these sciences track real causal patterns and enable us to make inductive inferences.

Kinds of natural kinds

5.1 INTRODUCTION

In previous chapters, I defended the existence of natural kinds in the special sciences, including the biological and social sciences, in fairly general terms. Although I have made reference to a number of properties and kinds drawn from a variety of sciences, from fluid mechanics to sociology, I have done so in the course of addressing general concerns about the possibility of natural kinds in these domains. In this chapter I will focus more closely on several case studies drawn from the physical, chemical, biological, physiological, and social sciences, in order to provide additional justification for some of the claims made in previous chapters as well as to clarify further certain features of natural kinds as they occur in various branches of science. These case studies will also serve to add detail to the account provided thus far (the simple causal theory of natural kinds) and will in some respects qualify and test the claims that I have made about natural kinds.

5.2 LITHIUM

Chemical elements are arguably the least controversial instances of natural kinds. In previous chapters, I made several references to the element *lithium* (atomic number 3) and used it to illustrate various points about natural kinds. In this section, I will repeat some of the features of this natural kind, elaborating on them and trying to draw some further conclusions about the structure of this natural kind, as well as the natural kinds corresponding to other chemical elements.

Atoms of lithium have one property, atomic number, which seems to give rise to a range of others. Like other elements of the periodic table, lithium is a particularly fertile natural kind, possessing projectible properties appearing in far-flung contexts and in diverse domains. Lithium is a silver-colored, soft metal, a solid at room temperature, with a density of 0.534 g cm^{-3}. It is an alkaline metal, highly corrosive, flammable, and a good conductor of heat and electricity; it reacts with water, and has a melting point of 180.5 °C, among many other properties. Moreover, its chemical properties make it suitable for use in lithium-ion batteries and as a mood-stabilizing drug due to its neurophysiological effects on the human brain. In both cases, what is crucial to the relevant chemical reactions is the tendency of lithium atoms to give up a single valence electron and become ionized. The resulting monovalent cations (Li$^+$) enter into chemical reactions that have far-reaching effects on macroscopic systems such as human brains and rechargeable batteries.

In addition to their chemical and biochemical properties, some atoms of lithium also have important nuclear properties, having to do with the radioactive decay of some lithium isotopes. As I mentioned before, atoms of lithium are not all identical to one another, since there are several different isotopes of lithium, two of which are stable (*lithium-6* and *lithium-7*), while several others are highly radioactive and have half-lives ranging from hundreds of milliseconds to something of the order of 10^{-24} s.[1] These isotopes differ not just in their half-lives but also in their patterns of decay. For example, *lithium-8* exhibits beta-minus decay while *lithium-4* decays by proton emission. Thus, they have very different causal properties and are implicated in different causal processes.

It may seem as though *lithium* is a paradigmatic example of a monothetic natural kind and that the necessary and sufficient condition for being a member of the natural kind *lithium* is possessing the property *atomic number 3*. Moreover, this one property would appear to be causally responsible for the instantiation of all other properties associated with this natural kind, from its basic physical properties (e.g., *has a melting point of 180.5 °C*), to its chemical properties (e.g., *reacts with water*), all the way to its neuropsychological properties (e.g., *acts as a mood-stabilizing drug*). Therefore, it may be tempting to say that membership in the natural kind *lithium* is strictly equivalent to having the property *atomic number 3*. However, most of the familiar properties of the natural kind *lithium* also depend on mass number in important ways. For instance, melting point is

---

[1] Some properties reported of lithium are composites and reflect the distribution of the two stable isotopes on the planet. For instance, when the atomic weight of lithium is given as 6.94, this reflects the fact that the vast majority of lithium on Earth is lithium-7 and the rest is lithium-6. This is not a property of a single lithium atom but of macroscopic samples of lithium as found on Earth. It is an average of the mass numbers of the two stable isotopes when their relative prevalence is taken into account.

generally a function of mass number rather than atomic number. More importantly, the chemical and other properties of lithium are contingent on the stability of the relevant isotope, since atoms of isotopes with extremely short half-lives cannot enter into chemical reactions (the fastest of which take place at a rate of the order of 10^{-15} s). Hence, most of the properties that I have listed concerning the element lithium apply only to the two stable isotopes, lithium-6 and lithium-7. In light of this information about its causal properties, what is the status of the natural kind *lithium*, and what is the relationship between the natural kind *lithium* and its associated properties? As far as I can tell, there would seem to be four options:

(1) There is no natural kind *lithium*, but rather natural kinds corresponding to each of the instantiated isotopes of *lithium* (from *lithium-4* to *lithium-12*). Some of these isotopes, the stable ones (namely, *lithium-6* and *lithium-7*), have most of the properties that we commonly associate with lithium, while the others do not. Each isotope is associated with a conjunction of two properties (atomic number and mass number) that are necessary and sufficient for membership in the kind.

(2) In addition to the natural kinds associated with each of the isotopes of *lithium*, as in (1) above, there is another kind, *lithium*, which corresponds to a simple disjunction of the natural kinds corresponding to all the instantiated isotopes (from *lithium-4* to *lithium-12*).

(3) The natural kind *lithium* has a necessary condition associated with it (*atomic number 3*) in addition to a disjunctive necessary condition comprising the two stable isotopes, leading to a cluster of properties, *atomic number 3 and (mass number 6 or mass number 7)*; the conjunction of these two conditions is both necessary and sufficient. Anything that has this complex property is a member of the kind *lithium*, and anything that lacks this property does not. There may also be natural kinds corresponding to each stable isotope.

(4) The natural kind *lithium* has a necessary condition associated with it (*atomic number 3*) in addition to a disjunctive necessary condition that comprises *all* the instantiated isotopes (*mass number 4... or mass number 12*), and the conjunction is both necessary and sufficient for something to belong to the natural kind *lithium*. Furthermore, the focal instances of kind membership are samples of isotopes with mass numbers 6 and 7, while the marginal instances have other mass numbers. There may also be natural kinds corresponding to each isotope.

Despite its apparent messiness, I would argue that the last option is the most consistent with philosophical principles and does the least violence to scientific practice. To see this, briefly consider each of the other options in turn. The first option simply does away with the natural kind *lithium*, and that appears to be a high price to pay, given that the chemical elements are perhaps the most widely acknowledged examples of natural kinds in the philosophical literature and play such a central role in science. The second option posits a disjunctive natural kind, which, I have argued, undermines projectibility. The third option would have it that unstable radioactive isotopes of lithium are simply not lithium, a notion which is not in keeping with scientific usage or ordinary parlance. Meanwhile, the fourth option is not disjunctive, since only one conjunct associated with the category is disjunctive (it is of the form P_1 & $(P_2 \vee P_3 \vee \ldots P_n)$), a fact which, as I argued in section 1.4, is compatible with projectibility. It allows for unstable isotopes to be members of the kind but it considers them marginal members, so as to account for the fact that they do not exhibit many of the other properties commonly associated with the kind *lithium*. This option therefore introduces graded membership into the kind *lithium*, which is a feature that is manifested in other paradigmatic natural kinds like some chemical compounds and biological species (see section 2.3). A similar argument can be made for virtually every other element in the periodic table. Hence, even the paradigmatic natural kinds, chemical elements, turn out not to be monothetic kinds identifiable with a single property or a simple conjunction of properties but polythetic kinds associated with a cluster of properties, and membership in the natural kind has also turned out to be graded.

One aspect of this picture that needs clarifying is the relationship between the primary properties and secondary properties associated with *lithium*. Generally speaking, for other natural kinds, I have suggested that this relationship is a causal one, the primary properties being causally responsible for the secondary ones. But some philosophers might balk at this suggestion when it comes to a natural kind like lithium, since the loose cluster of primary properties, as in (4) above, serve, rather, to *determine* the macroscopic properties of lithium, like melting point and reactivity with water. In similar contexts, the microproperties are also sometimes said to be the *realizers* for the microproperties. This relation of determination or realization is often thought to be a *sui generis* relation that obtains between the micro- and macrolevels and is distinct from an ordinary causal relationship. But this does not seem obvious on at least some conceptions of causation. Consider the melting point of lithium, which relates primarily to mass number rather than atomic number. Melting a solid involves a transition from solid to liquid phases entailing a change in the nature of the chemical bonds. In an element like lithium, this takes place when a collection of atoms are bonded together in crystalline form. The properties of these atoms when in a crystalline state, including mass number, can be said to be causally implicated in the process of melting, which is a process that has temporal duration and takes place as a result of a change in ambient conditions, primarily temperature. There does not seem to be any obstacle in regarding the microproperties of lithium as being at least partly causally responsible for its macroproperties. Many of these causal properties are also dispositional, requiring certain other conditions to be in place for their manifestation (for example, temperature for melting point).

There are two further features of this natural kind that are not usually associated with natural kinds in the basic sciences that need to be emphasized. First, I have claimed that on the most plausible understanding of the cluster of properties that are associated with the natural kind *lithium*, some combinations of these properties yield members of the kind that possess more of the other properties associated with lithium. Hence, it is warranted to think of those members as being focal members of the kind, while the others can be regarded as marginal members. This finding of graded membership in the kind suggests that this natural kind can be represented as being associated with a weighted cluster of properties (along the lines suggested in section 2.4), with focal members possessing more of the secondary properties associated with the kind. In this case, the fact that there is graded membership in the kind *lithium* does not mean that there are members that are intermediate between lithium and some other chemical element. Rather, the kind "trails off" with highly radioactive isotopes being marginal members of the kind. The kind can be considered a fuzzy kind, but in this case fuzziness does not imply that there are members intermediate between it and some other kind, but just that there are focal and marginal members of a single kind.[2] The other notable feature of the natural kind *lithium* is that it is involved in crosscutting systems of natural kinds (as already mentioned in section 3.6). *Lithium* has subkinds corresponding to its various isotopes, and some of these subkinds

[2] Sober (1980) has actually suggested that the former kind of fuzziness may obtain for atomic elements. After describing the nuclear reaction in which nitrogen is transformed into oxygen (absorbing an alpha particle and emitting a proton), Sober (1980, 357) asks: "At what point does the bombarded nucleus cease to be a nitrogen nucleus and when does it start being a nucleus of oxygen?" But I think that it is harder to make the case that these transient states of atomic nuclei are intermediate members of the two kinds *nitrogen* and *oxygen*.

can also be classified with isobars of other elements (e.g., *lithium-8* with *helium-8*). More significantly, both *lithium-8* and *beryllium-8* can be categorized as *beta-minus decay nuclides*, which exhibit a characteristic pattern of radioactive decay by emitting an electron from the nucleus.

5.3 POLYMER

An interesting candidate for a natural kind is the chemical category *polymer*, which has a subbranch of science devoted to it (polymer science). Polymers defy some of the philosophical assumptions commonly made about kinds in the natural or basic sciences – in this case, chemistry. Polymers are long chains of molecules, often in regular repetitive patterns, consisting of one or more smaller units (monomers). Some are composed of a single monomer that repeats indefinitely, often thousands of times, and sometimes billions of times in a single molecule, while others are composed of a few such units that repeat either regularly or irregularly in a chain. Some of these substances occur naturally (e.g., rubber, cellulose, silk) while others are manufactured (e.g., polyethylene, polystyrene). The best-known polymeric molecule, DNA, is composed of four smaller monomeric units, which repeat irregularly in the famed double helical strand.

Chemical kinds are often assumed to be the archetypal natural kinds, identified on the basis of the composition of their most basic (identical) microphysical elements, from which they derive their macrophysical properties.[3] The kind *polymer* disrupts this assumption in a number of ways. First, the category *polymer* itself obviously does not classify substances on the basis of their composition, since polymers are quite diverse in terms of the components that make them up. *Polymer* does not denote a type of compound that possesses identical microphysical elements. Rather, what polymers in general have in common is that they consist of long, chain-like molecules, which often involve carbon or silicon, but also other elements. Moreover, many of the properties common to all polymers are related in some way to the properties of the chains, their linking and cross-linking patterns, and the ways in which they behave at different temperatures and under other conditions. *Polymer* is not a microstructural kind in the straightforward sense, but nor is it clearly a functional kind since it pertains explicitly to the structure of a compound, not its function.

---

[3] In fact, Needham (2000) shows that even the chemical compound *water*, along with many other chemical compounds, does not conform to this simplistic picture of identical microconstituents.

Second, the molecules that compose each specific polymeric substance are generally not identical to one another. They can vary in length dramatically, some polymeric molecules of a single substance being much longer than others. Copolymers, which are composed of more than one recurring monomeric unit may not even contain these monomers in the same sequence. Some copolymers are repetitive (e.g., A-B-A-B-A-B...) while others have their monomers arranged in random patterns (e.g., A-A-A-B-A-B-B-A-A...). The latter, termed *random* or *statistical copolymers*, usually contain a fixed proportion of each monomer.[4] Third, what determines the properties of any given polymeric substance is not solely the monomeric units that are the building blocks of the polymeric molecules. Even when a polymeric molecule consists of a single repeating monomeric unit (e.g., A-A-A-A...), the properties of the polymeric substance are generally very different from that of the substance that is constituted by a collection of unattached monomeric molecules. For instance, the properties of polyethylene are entirely different from those of ethylene, even though polyethylene consists of long chains of ethylene monomers linked together by covalent chemical bonds.

A skeptic might deny that the category *polymer* corresponds to a natural kind since all that polymers seem to have in common is that they happen to consist of long chains of molecules with repeating units. If there were nothing that all polymers shared aside from being composed of macromolecules, then they would arguably not constitute a natural kind. There might be no greater claim to this category picking out a natural kind than there is for thinking that *tall person* constitutes a natural kind, or that *large mammal* does. Even if members of the category *polymer* share a single property or are similar in a certain respect, that is not enough to consider all polymeric substances instances of a natural kind. As I emphasized in Chapter 2, to understand the natural world it is not enough to take note of similarities but we need to identify similarities that depend on shared causal properties. Such "important" properties (to use Mill's expression), which are causally linked to yet other properties, should enable us to explain and predict natural (and social) phenomena. This is certainly the case here, since polymers as such share numerous causal properties that are relatively independent of the types of atoms that make them up. Some of

---

[4] The most famous of these statistical copolymers is, of course, DNA, which contains four different monomers, and though the pattern in which they are arranged in each strand of DNA is in some sense random, the distinctive pattern of each strand is crucial to the biochemical properties of the molecule.

these properties follow in a direct causal fashion from the fact that they consist of large macromolecules with repeating monomeric units (call this property, P_1), such as the fact that polymer molecules tend to pack together in a nonuniform pattern (P_2). There are also numerous other causal properties of polymers as such. Unlike other chemical substances, polymers do not enter into a perfect solid (or crystalline) phase, since different parts of each molecule crystallize independently. Instead, they are usually characterized by a semicrystalline phase (P_3), though some are also noncrystallizable because there is no long-range order in the positions of the molecules that constitute them. The semicrystalline state of most polymers is associated with an increase in rigidity (P_4) and tensile strength (P_5). In addition, polymers are generally characterized by high viscosity (P_6), for reasons having to do with the resistance to flow of substances characterized by extremely long molecular lengths (recall some of the observations made about the property of *viscosity* in Chapter 3). For similar reasons, polymers are not found in gaseous form (P_7).[5]

Hence, polymers are characterized by a necessary property (P_1) – namely, their consisting of long molecular chains consisting of one or a few repeating monomers. From this primary property flow various others as a direct causal result. Many of these other properties do not apply strictly to all polymers without fail, but they apply sufficiently widely as to enable us to make reliable inductive inferences. As we would expect when it comes to many kinds in the special sciences, the causal links between the properties associated with the kind are not all strict but probable or statistical. Accordingly, the causal laws involving polymers are not exceptionless generalizations but may contain *ceteris paribus* clauses. Moreover, some of the properties associated with polymers are complex functions of other properties, such as the property of increasing in rigidity when entering the semicrystalline phase. Most importantly, perhaps, the properties associated with polymers are not all on a par, since some of them are causally involved in the production of other properties. Instead of a picture whereby kind K is simply associated with a set of properties, $\{P_1, P_2, P_3, P_4, P_5, P_6, P_7\}$, we have a model according to which one property, P_1, is primary, in the sense that it is causally efficacious in instantiating P_2 (long molecular chains tend not to pack together in a uniform pattern), which is, in turn, causally efficacious in producing P_3 (nonuniformly packed molecules tend not to have a crystalline phase but

[5] Here and elsewhere in this section, I am relying primarily on Grosberg and Khokhlov (1997), as well as on Gratzer (2009).

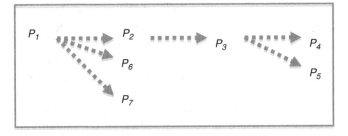

Figure 5.1. Causal relationships among some of the properties
associated with the kind *polymer*

enter into a semicrystalline phase), which then leads to P_4 (semicrystalline macromolecules show increased rigidity) and P_5 (increased tensile strength). Additionally, the primary property (P_1) tends to lead directly to other properties – for instance, to high viscosity (P_6) and to the lack of a gaseous phase (P_7). In Figure 5.1, I have tried to diagram these causal relationships among some of the important properties associated with the kind *polymer*, with dashed arrows indicating probabilistic (rather than strict) causal links. Needless to say, the properties that I have picked out represent only a portion of the properties associated with the kind *polymer*. If one were to track all the causal properties that polymer scientists, organic chemists, biochemists, materials scientists, and others have discovered of polymers, then it would resemble a large causal network consisting of multiple branching patterns. The natural kind *polymer* with its associated primary property (P_1) is what gives rise to this causal cascade, generating various real causal patterns.

One question that emerges at this point concerns the role of what I have been calling the primary causal property of polymers (P_1) – namely, their being composed of large molecular chains consisting of repetitive monomeric units. Since this property is a necessary property, which distinguishes this kind from others, it may appear that the kind *polymer* is simply identical with this property. In this case and in at least some others (recall the kind *Newtonian fluid* discussed in Chapter 3), there is one property that plays a central role, in the sense that it is necessary for membership in the kind and is causally efficacious in giving rise to other properties associated with the kind. Hence, it does seem as though there are some natural kinds that are primarily associated with a single property. The philosophical tradition has considered natural kinds to be associated with clusters of properties, but at least in some cases it appears as though one property is primary and that it gives rise to others, which are arranged in the form of a causal network as opposed to a mere cluster. However, it is still possible that our standards for what constitute natural *properties* may be different from what constitute natural *kinds*, since we expect a natural kind to play an important role in a causal network, leading to the instantiation of many other properties, whereas each property in the network need not itself be central to a causal network of its own (this point will be further justified in section 6.2). In fact, some of these properties will play a distinctive causal role in this context only in conjunction with yet other properties associated with the particular kind in question. For instance, the property of *high viscosity* plays a characteristic causal role in the context of large macromolecules, leading to effects different from those that it generates in other contexts (such causal relationships, linking different branches of the causal network, are not shown in Figure 5.1).

A more substantive issue arises when one observes that, in this case at least, the primary property is not only necessary for membership in the kind but also is sufficient. For not only do all polymers consist of large molecules composed of repetitive monomeric units, but also any substance consisting of such molecules is a polymer. If so, this may now suggest a vindication of the essentialist view that each natural kind is associated with a property (or properties) that is both necessary and sufficient for membership in the kind. Although this may well be a case in which there is a single property that is both necessary and sufficient for membership in a kind, that should not amount to an endorsement of essentialism, since there are a number of other conditions that essentialists impose upon kinds, as discussed in Chapter 1 (which I have argued are unwarranted). Moreover, in this case, if P_1 is indeed a property that is both necessary and sufficient for membership in the kind *polymer*, as it appears to be given our current state of knowledge, it is a property of a peculiar sort. As I have already pointed out, it is not a microphysical property in the standard sense, since all polymers do not consist of identical microphysical components. Perhaps more importantly, the property is inherently vague, since it does not specify *how large* polymeric molecules need to be to qualify as large molecules or macromolecules. There is no easy answer to the question concerning the requisite size of polymeric molecules, and this fact renders the boundary of this natural kind a fuzzy one. *Polymer* would appear to be yet another example of a bona fide fuzzy natural kind. In this case, the fuzziness arises as a result of the fact that the main property associated with polymers is a continuous or dimensional one and that there is no sharp cutoff between members and nonmembers. There is also room here for thinking of membership in the natural kind as being graded, with some compounds being focal examples of polymers and others more marginal ones (the category *oligomer* is sometimes used to classify substances with molecules intermediate between monomers and polymers).

There is one last feature of this natural kind that both reinforces earlier conclusions and illuminates them in significant ways. Here as elsewhere, the kind *polymer* can be divided into subkinds, each of which has additional properties in common not shared by the class of polymers as a whole. I have already mentioned the difference between polymers consisting of a single repetitive monomeric unit (*homopolymers*) and those that consist of more than one such unit (*copolymers* or *heteropolymers*). When we conjoin the primary property that characterizes all polymers with each of these additional properties, we generate yet further properties. Hence these can be considered natural kinds in their own right, subkinds of the natural kind *polymer*. In some cases, the causal patterns generated by subkinds can be represented on Figure 5.1, which captures one aspect of the causal relations involving polymers. For instance, those polymers that do not enter into a semicrystalline phase are noncrystallizable polymers. At this point, the causal network can be thought to encounter a fork and can be split into two new causal networks, each corresponding to a distinct subkind of polymer: *semicrystallizable polymers* and *noncrystallizable polymers*. But in other cases, the different subkinds are generated by considering an entirely different aspect of polymer behavior – for example, whether they consist of a single repeating monomeric unit or more than one (*copolymers* and *heteropolymers*). Since this distinction captures a causally independent aspect of polymer behavior, the kinds into which we divide polymers can crosscut one another (*co-* vs. *heteropolymers, semicrystallizable* vs. *noncrystallizable polymers*, and so on). This is a further illustration of the phenomenon of crosscutting kinds discussed in section 3.5, reinforcing its *aspectual* nature. Depending on whether we are interested in the phase changes that polymers participate in or the number of units from which their molecules are composed, we will divide the kind *polymer* into distinct, crosscutting subkinds, each implicated in its own set of causal processes or causal patterns.

Several points have emerged from this discussion of the natural kind *polymer*, some of which have already been adumbrated in earlier chapters. As in the case of the natural kind *lithium*, the properties associated with a natural kind are generally not all on a par, since some are causally efficacious in producing others. Rather than a mere cluster of properties, natural kinds are associated with a causal network of properties, representing the causal interactions in which the natural kind can participate. Though the kind *polymer* is a monothetic kind associated with a single necessary and sufficient property, that property is a dimensional one that gives rise to a fuzzy kind with graded membership.

5.4 VIRUS

Another promising candidate for a natural kind, this time drawn from the field of biochemistry or microbiology, is the category *virus*, which is also the subject of a subdiscipline of science (virology). The question here again is whether the category *virus* corresponds to a natural kind, and if so, what makes it such, and whether there are subcategories of the category *virus* that are themselves natural kinds. A virus has been famously characterized by Peter Medawar as "a piece of bad news wrapped in protein," the bad news being a strand of genetic material (either DNA or RNA), which in most cases ranges from just 1 to 15 genes or so. Deceptively simple, viruses seem to have a winning formula for survival and reproduction. Many virologists do not consider them to be living organisms, since they cannot replicate themselves unless they can find a suitable cell in a host organism. But even though they cannot reproduce by themselves, virus particles are remarkably successful at doing so provided they can find the appropriate cellular host. Once a virus particle (or "virion") finds the appropriate host cell, its genetic material enters the cell, carrying instructions for the construction of new virions by assembling the appropriate building blocks from that cell. Then, each new virion, conveying more or less the same genetic material wrapped inside a protein casing, transports that genetic material to another host cell to carry out the same process.[6]

As this brief account shows, there would seem to be a (rather weak) necessary condition on something belonging to the kind *virus*: the property of having or being a nucleic acid genome (DNA or RNA) contained in a proteinaceous particle. But this necessary condition does not appear to be sufficient, since one could easily conceive of a strand of DNA wrapped in a protein packet that is incapable of initiating the *infectious cycle* that is so characteristic of the viral *modus operandi*. In terms of Medawar's ironic description, the "bad news" needs to be in a form that allows it to be transmitted, broadcast, and reproduced. Hence, the causal process that characterizes the life cycle of a virus is central to the understanding of what it is to belong to the kind *virus*. But the virus life cycle is one that is

---

[6] For more on viruses, see Crawford (2011) and (2000), as well as the valuable internet resource: www.virology.ws/.

couched in terms of what is *typically* the case. There are few universal generalizations about viruses though there are a slew of nonstrict causal statements that characterize what usually happens in the process by which viruses reproduce themselves. There are some constant features to be sure, principally that viral genomes must make messenger RNA (mRNA) that can be read by the ribosome of the host cell. But the way in which they go about making the mRNA depends on the nature of the genetic material they have to begin with – for instance, whether it is single-stranded DNA (ssDNA), double-stranded DNA (dsDNA), single-stranded RNA (ssRNA), and so on. According to the current understanding, there are seven variants on the type of genetic material carried by a virus, each of which involves a different strategy for producing mRNA. After the initial characterization of a virus as a protein particle containing a genome (P_I), which can be considered a necessary property associated with the kind *virus*, there is another necessary property – namely, that the viral genome is capable of making mRNA that is readable by the ribosome of a host cell (P_2). This second property is highly dispositional and makes reference to two other kinds (*mRNA*, *ribosome*), and it does not seem possible to specify the other necessary property for membership in the kind *virus* in any way that is not dispositional. If these two properties are in place, there are a number of causal properties that follow, some of which may be different depending on which of the seven types of genetic material are involved, but each of which corresponds to a distinct subkind of virus.

The two principal properties of viruses, P_I and P_2, are jointly causally responsible for a number of other properties, such as those associated with the *infectious cycle* of a virus, which is of course dependent on finding a *susceptible* and *permissive* host cell (respectively, a cell that has a receptor for a given virus and one that is capable of replicating the virus). If such a host cell is found, the virus is able to attach itself to the cell (P_3), the viral mRNA is translated by host ribosomes (P_4), the viral genome is replicated (P_5) and assembled (P_6), and the particles containing the genome are released by the cell to infect other cells (P_7). The process is repeatable indefinitely and the virus is causally responsible for each stage of the process. The stages may not be what we might ordinarily consider properties of the virus, but they are indeed dispositional properties or capacities that are possessed by every virus, and they depend both on the configuration of its protein particle (which enables it to attach to the host cell) and the makeup of its genetic material (which enables it to be replicated by it), as well as on the nature of the host cell. This causal process is characteristic of all viruses and is common to them all, no matter the specific makeup of their genetic material. The common and repeatable causal process enables virologists to make empirical generalizations about viruses and to project from viruses already observed to ones that have not been. This is what makes *virus* another example of an apparent natural kind.

Beyond the common causal process initiated by all viruses, there are more specific causal processes that are initiated by only some viruses. These are specific to subkinds of the kind *virus*. We have already seen that there are seven basic types of viral genomes (ssDNA, dsDNA, etc.) leading to seven broad kinds of virus, which is the basis of the Baltimore classification system of viruses. This classificatory system relates to the process whereby mRNA is produced: Some viruses begin with a single-stranded DNA molecule, others with a single-stranded RNA molecule, and so on, but all end up producing mRNA to direct the synthesis of proteins within the host cell. Hence, this sevenfold taxonomy classifies viruses on the basis of the process that leads to the production of mRNA, which is necessary for viral reproduction. Each type of virus takes a different causal route to reproducing itself and these differences, associated with repeatable causal processes, are associated with genuinely different kinds of viruses.

The sevenfold classification of viruses, based on the type of genetic material they contain, is not the only way in which subkinds of viruses can be identified. Viruses may also be subdivided by other criteria. The main competitor to the Baltimore classification is one based roughly on the Linnaean system in biology, which attempts to classify viruses into the traditional phylogenetic taxonomic categories (species, genus, family, order, and so on) based primarily on descent. This classification scheme crosscuts the Baltimore scheme, since closely related viruses sometimes have different types of genetic material (i.e., DNA, RNA, etc.), and conversely, viruses that are not so closely related may use the same type of genetic material to produce mRNA. As with the phylogenetic classification system for other organisms, it is sometimes hard to establish a correct taxonomy because of the difficulty of determining relationships of descent. The problem is compounded with viruses because of the lack of a fossil record. Nevertheless, this method of classification is widely used in the classification of viruses, notably by the International Committee on Taxonomy of Viruses (ICTV). But even though many virus taxonomists conceive of themselves as attempting to delineate relationships of descent, the principal method for discovering such relationships lies in sequencing the genomes of viruses. It may be tempting to think that any two virus particles that share the same (or largely the same) genetic material belong to the same subkind or virus species. The problem is that when virus particles replicate, their genotype and phenotype can vary widely, and that is indeed one key to their success. These extreme fluctuations make it rather difficult to identify a species of virus with a particular type of genetic material or even a range of genomes. That is why the phylogenetic system faces particular challenges when it comes to virus classification. When it comes to virus *species*, the ICTV explicitly defines it in ecological rather than phylogenetic terms: "Virus species is a polythetic class of viruses that constitute a replicating lineage and occupy a particular ecological niche."[7]

Some attempts to subdivide the kind virus are based on etiology or causal history while others are based on synchronic causal properties. But what of the kind *virus* itself? So far, I have understood the kind *virus* to be identified on the basis of its synchronic causal properties or the causal processes that it initiates, but as we have already seen (section 4.3), many biological kinds are individuated according to etiology or causal history. Should the kind *virus* be understood as identifying a type of entity with a common causal history? Do all viruses not share the same origin and line of descent as do, for example, members of a biological species? One problem with such a proposal is the fact that the origin of viruses is shrouded in mystery, with one theory hypothesizing that they originated from simple genetic elements before the evolution of cells and another theory positing that they evolved from parasitic unicellular organisms that lost their cellular structure within the host cell. Some virologists even argue that viruses have multiple independent origins (Moreira and López-García 2009). If so, then *virus* would not be an etiological kind at all. Even if it were one, what primarily unites viruses into a single kind would appear to be a set of synchronic causal properties rather than etiology or causal history.

In the case of viruses, there are a few necessary conditions in the form of causal properties that pertain to the kind as a whole, properties which then lead members of the kind to enter into causal interactions of a uniform or very similar character, enabling us to make a range of loose generalizations and projections concerning viruses. We have also found that there are credible subkinds of the natural kind *virus*, and indeed, that systems of subkinds can crosscut one another, based on the different types of causal process that they enter into or their different causal histories. Finally, the natural kind

---

[7] www.ictvonline.org/virusTaxInfo.asp. It is impossible here to delve into the "species problem" or the debate about whether species are natural kinds. For a collection of classic papers on the topic, see Ereshefsky (1992), including Van Valen (1976/1992), which first proposed the ecological species concept. For more recent philosophical essays, see Wilson (1999).

virus itself is individuated primarily on the basis of synchronic causal properties rather than etiology or causal history.

5.5 CANCER AND CANCER CELL

It is not uncommon to read that, "Cancer is not a single disease with a single cause" (Crawford 2000, 155), and cancer is sometimes referred to as a "family" of diseases. Indeed, the history of cancer is sometimes written in such a way as to imply that ancient writers were mistaken to lump all cancers together and that modern medicine has split cancer into a number of separate kinds with distinct causes. This conventional wisdom would have it not that *cancer* is a natural kind with distinctive subkinds, but that it is not a single natural kind at all. But even though cancer does not have a single cause (and hence is not an etiological kind), cancer itself is a causal process with a set of common features. In this section, I will argue that the category *cancer*, which pertains to physiology, medicine, and related sciences, denotes a natural kind of process. Cancer, like many other diseases, refers to a kind of spatiotemporal process that may be instantiated in individual multicellular organisms. In previous sections, the natural kinds that we have been concerned with have generally been kinds of entities or individuals rather than kinds of process, though most of them have also been implicated in causal processes in which some of their dispositional properties are manifested. In this case, the kind of entity that is most closely implicated with the process of *cancer* is *cancer cell*. I will argue that once *cancer cell* is understood as a natural kind of entity, then *cancer* can be seen as a natural kind of process.

In the past couple of decades, evidence has been accumulating for a unitary treatment of cancer. Indeed, in recent work, some cancer researchers have posited that cancer has six "hallmarks," each of which marks a stage in the development of the disease (Hanahan and Weinberg 2000, 2011). But these "hallmarks" of cancer are more accurately described as causal properties of the *cancer cell*. "Cancer cells are the foundation of the disease; they initiate tumors and drive tumor progression forward, carrying the oncogenic and tumor suppressor mutations that define cancer as a genetic disease" (Hanahan and Weinberg 2011, 661). The *cancer cell* is, I will argue, a natural kind based on its causal properties, in much the same way as some of the other categories examined in this chapter and elsewhere in this book.

The hallmarks of cancer cells, according to this research program, are the result of mutations that occur in the genomes of these cells, resulting in mutated "oncogenes." What kinds of genetic mutations give rise to cancer cells? The six properties of cancer cells delineated by Hanahan and Weinberg (2011) are as follows. First, *sustaining proliferative signaling*: Due to genetic mutations, cancerous cells deregulate the signals that control cell growth and cell division, hence unleashing a cycle of unrestrained growth and division relative to normal cells. There are a number of genes that carefully regulate cell division, restricting it to specific stages in the life cycle of a cell, but mutations in at least some of these genes can result in uncontrolled proliferation of cells. Second, *evading growth suppressors*: Cancer cells also contain mutations that disable genes that suppress cell growth and proliferation. There is considerable redundancy in the mechanisms of proliferation suppression, meaning that cancer cells must have multiple ways of evading these mechanisms. One relatively straightforward way in which they do so is through the inhibition of cell contact, since contact between cells normally has the effect of suppressing cell growth. Thus, cancer cells have various mechanisms of contact inhibition, again due to mutations in their genotype. Third, *resisting cell death*: Another thing that cancer cells do is that they do not die as normal cells are programmed to do after a certain length of time. Cancer cells have mutations that interfere with programmed cell death (apoptosis) and so prolong their lives. Fourth, *enabling replicative immortality*: Cancer cells also acquire mutations that enable unlimited replicative potential, and they do so in various ways, notably by lengthening telomeric DNA in such a way as to prevent senescence of cells; that is, preventing them from entering a viable but nonreplicative state, as normal cells do. Fifth, *inducing angiogenesis*: Cancer cells need blood vessels to nourish them and provide oxygen, as well as to carry away waste and carbon dioxide, and angiogenesis is the process of growing new blood vessels to carry out these functions. To survive and flourish, cancer cells promote this process, which is usually switched off in adult mammals (except temporarily in healing wounds). This operation is governed by genes manufacturing signaling proteins that orchestrate the growth of new blood vessels, which are reactivated in cancer cell genomes. Sixth, *activating invasion and metastasis*: To enable migration to other areas of the body, cancer cells undergo various changes, such as alterations in shape and in their attachment to other cells. One type of genetic mutation facilitating this operation involves suppressing the production of cell-to-cell adhesion molecules, which allows cancer cells to migrate to other areas of the body.

This characterization of cancer cells in terms of six key properties raises the following question: Why do these "hallmark" properties happen to be co-instantiated in certain cells? Though the mechanisms that lead to some of these properties may not be entirely independent of all the others,[8] others seem to be autonomous, so the co-occurrence of the requisite genetic mutations might appear to be something of a mystery. Since genetic mutation is ordinarily rare in the somatic cells of an organism, there is yet another class of mutations that allow the mutations listed above to occur and persist. These "enabling" mutations involve "caretaker genes," including "those whose products are involved in (1) detecting DNA damage and activating the repair machinery, (2) directly repairing damaged DNA, and (3) inactivating or intercepting mutagenic molecules before they have damaged the DNA" (Hanahan and Weinberg 2011, 658). It is not a coincidence that the distinct genes responsible for the six "hallmark" properties enumerated above are found in mutant forms in cancer cells, since the mutations in the parts of the genome that normally repair mutant genes are the ones that enable the occurrence of these other mutations, or rather ensure that they are not corrected. Hanahan and Weinberg (2011, 658) dub this property "an enabling characteristic," since mutations in the "caretaker" genes are a necessary condition for the persistence of the other mutations in cancer cells. When these mutations in caretaker genes occur, they can be transmitted to the next generation of cells, thus enabling other mutations to ramify, resulting in fully cancerous cells with the properties enumerated above. When cells acquire one or more of the six "hallmark" properties, these properties can then be passed on to the next generation of cancer cells uncorrected. The mutations that enable cells to acquire these properties are the ones that ensure the spread of these mutated cells. It is not a coincidence that these properties are co-instantiated, since without them, or at least some of them, cancer cells would not enjoy the success that they do and might be weeded out. Cancer cells with favorable mutations multiply and outgrow their competitors in a version of natural selection that takes place, not among organisms, but among cells within an individual organism. Of all the natural kinds we have looked at, this is perhaps the one most in line with Boyd's homeostatic property cluster account (section 2.6). In this case, the mutated

---

[8] For instance, Hanahan and Weinberg (2011, 655) observe: "One important implication, still untested, is that the ability to negotiate most of the steps of the invasion-metastasis cascade may be acquired in certain tumors without the requirement that the associated cancer cells undergo additional mutations beyond those that were needed for primary tumor formation." Note also that some of the hallmark properties of cancer can be attributed to mutations while others may be due to epigenetic mechanisms such as gene methylation, which is the process whereby genes are switched on and off (Hanahan and Weinberg 2011, 658).

caretaker genes can be thought of as the causal mechanism that gives rise to the cluster of other properties associated with cancer cells. But even here, the other properties are not really kept in a state of homeostasis, since cancer cells typically mutate rapidly and acquire new properties.

Another question that arises is this: Which of these "hallmark" properties are necessary (and sufficient) for a cell to belong to the kind *cancer cell*? Recent research asserts that the mutations in the caretaker genes (P_1) are necessary for the other mutations to persist and accumulate, and hence for a cell to become genuinely cancerous. But P_1 cannot be considered the defining property of the category *cancer cell*, since even though it constitutes a necessary condition, it is far from sufficient. Genes with only this mutation are not cancerous unless they acquire at least some of the rest. As for the other properties (label them P_2–P_7), the taxonomic practice of oncologists suggests that there is no unique set of sufficient conditions. Rather than being a monothetic category, it would appear that *cancer cell* is a polythetic one, wherein one property is necessary (mutations in the caretaker genes), and this property in conjunction with at least some of the others is sufficient (as shown in Figure 5.2). In other words, *cancer cell* is a polythetic or cluster kind: One property provides a necessary condition for membership in the kind, and a set of other properties are such that at least some of them must be conjoined with the necessary property to generate a sufficient condition for membership. Unlike some of the other cases discussed in this chapter, in which a few necessary conditions are also sufficient for the characterization of the kind (*polymer*, *virus*), in this case the necessary condition is not sufficient on its own for membership in the kind. That is because cells do not become genuinely cancerous unless they also accumulate the mutations that give rise to the "hallmark" properties of cancer cells. These hallmark properties of cancer cells lead, in turn, to the causal process of cancer, which is a disease that interferes with the normal functioning of the organism. The causal process of cancer takes hold once the mutant cells, which have at least some of the properties enumerated earlier, proliferate without limit, grow uncontrollably, survive indefinitely, and so on. Without some of these additional properties, there would not be interference with the normal functioning of the organism. To see this more clearly and further justify this claim, the kind *cancer* needs to be considered as well.

A standard textbook on cancer biology characterizes cancer as follows:

Cancer is an abnormal growth of cells caused by multiple changes in gene expression leading to dysregulated balance of cell proliferation and cell death

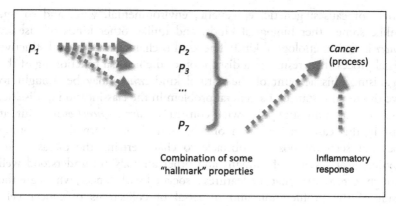

Figure 5.2. Causal relationships among the properties associated with
the kind *cancer cell*

and ultimately evolving into a population of cells that can invade tissues and
metastasize to distant sites, causing significant morbidity and, if untreated, death
of the host. (Ruddon 2007, 4)

When mutations occur that produce cancer cells with the "enabling" and
"hallmark" properties discussed above, leading them to mutate, replicate,
proliferate, and grow, this is likely to generate the cascade of events that
characterizes cancer. The characteristic properties of cancer ensue, such as
the formation of populations of cells that interfere locally with bodily
functions and then invade other parts of the organism. Moreover, recent
research indicates that the effects of the cancer cells themselves are compounded by the contribution of the organism in the form of the body's
inflammatory response. Paradoxically, the body's natural response to a
perceived threat (which includes cancer cells or collections of such cells)
facilitates the proliferation of cancer cells and the growth of tumors. That
is because the immune system is also involved in healing wounds, among
other functions, and cells dedicated to these functions can be recruited in
nourishing cancer cells. Hence, when we speak of cancer as a disease or a
natural kind of process, the properties of the cancer cell must be conjoined
with the body's inflammatory response to characterize the causal process
characteristic of cancer (see Figure 5.2). I will not try to elaborate on this
causal process any further, since its main features follow more or less
directly from the "hallmark" properties of the cancer cell, which have
already been outlined.

We saw earlier that there is no single cause for the mutations characteristic of cancer, since the mutations involved can arise as a result of a diverse

variety of causes: genetic, epigenetic, environmental, viral, and so on. Unlike some other biological kinds, and unlike other kinds of disease, *cancer* is not an etiological kind. The kind is characterized by a distinctive causal process that results in a disruption of the normal functioning of the organism. This account of the natural kind *cancer* may be thought to give rise to an instance of a general problem in the classification of disease, since it raises a question as to what constitutes *normal functioning*. But, at least in this case, the question of what constitutes normal functioning does not seem to pose an obstacle to characterizing the disease. The physiological systems of humans and other animals are understood well enough to recognize that the features associated with cancer, which are the result of the proliferation and dispersal of collections of cancer cells (tumors) in the body, are disruptive of normal functioning, often leading to morbidity. Following a number of philosophers of medicine, I would maintain that the relevant notion of normalcy can be articulated without commitment to a *nonbiological* notion of normativity or value-ladenness.[9] Whether or not a physiological process is cancer is not relative, say, to the moral or pragmatic values that we hold. The relevant notion of normalcy can be captured in biological terms having to do with the functions of the organism and its systems, and is not dependent on other values. However, normalcy *is* a matter of degree. There are intermediate cases between normality and abnormality, and there are processes involving cancer cells that permit many bodily processes to continue though they may impede them to a significant extent. But this fuzziness in the demarcation line between members and nonmembers of the kind *cancer* is a feature that I have already found in other natural kinds. In this case, the reason for the fuzziness has to do with indeterminacy in specifying which combinations of causal properties are sufficient to give rise to the natural kind of causal process, *cancer*. In other words, it has to do with the extent to which these combinations impede normal biological function, or the degree to which normalcy needs to be disrupted for a biological process to be considered a disease.

Since the "hallmark" properties of cancer issue directly from mutations in the genome of cancer cells, it may be tempting to think that a requisite set of mutated genes can be used to specify membership in the kind *cancer cell*. Some of these characteristic mutations have already been intensively studied by oncologists and seem to be implicated in a wide variety of cancers, such as the mutated versions of the oncogenes, *src*, *ras*, and *myc*,

[9] For recent discussions of medical kinds in this vein, see, e.g., Benditt (2010) and Williams (2011).

among others.[10] Other such genes are continually being identified and efforts are underway to identify all the mutations in the human genome associated with various types of cancerous cells, leading to a comprehensive genome of cancer. Some types of cancer cell contain a total of 50–80 mutated genes, while others contain only 5–10 (Mukherjee 2010, 451–452). But it is unlikely that there will be a requisite combination (or set of combinations) of mutated oncogenes that will be found to determine membership in the kind *cancer cell*. There would seem to be little prospect of specifying a set of genes that is necessary and sufficient for membership in the kind *cancer cell*, since what determines membership in the kind are the effects that result from the co-instantiation of the relevant properties, and these effects are vastly multiply realizable in terms of combinations of mutated oncogenes. As Mukherjee (2010, 453–454) notes, what is crucial is not the mutated genes themselves but the changes brought about in the physiological *pathways*, those responsible for cell division, growth, and related functions, each of which may be governed by multiple genes. As with other broadly functional kinds, there is no prospect of identifying them with a set of microstructures. We may not be able to anticipate the various different mutations and their combinations that would result in cells with the hallmark properties of cancer cells.

Cancer is commonly divided into subkinds that pertain to the organ or organ system that is the primary site of the tumor or cancer cells – for example, *lung cancer, breast cancer, prostate cancer, leukemia*, and so on. But many of these subordinate categories may not be true natural kinds, since they do not seem to share causal properties (beyond those associated with *cancer* in general), nor do they share a causal history, since lung cancer, for example, can be caused by a wide range of causal factors. At the level of the cell, there does not seem to be anything that all lung cancers have in common apart from the fact that they are cancerous and originate in lung tissue (most originate in epithelial cells of the lung tissue but some do not). *Lung cancer* is divided into two subcategories, depending on the overall appearance of the cancer cells, *small cell lung carcinoma* and *nonsmall cell lung carcinoma*, and the latter is subdivided into *squamous cell lung carcinoma, large cell lung carcinoma*, and *lung adenocarcinoma*. These subdivisions may be better candidates for natural kinds, since the cells responsible for each share some common properties, and the type of process associated with each also shares causal properties. But some of these subcategories also seem heterogeneous in terms of their associated properties, particularly

---

[10] See Mukherjee (2010, 357–383) for an account of the discovery of some of these genes.

large cell lung carcinoma, which seems to be an undifferentiated "none-of-the-above" type category. The current state of scientific knowledge does not appear to have delineated subtypes of cancer that are clear candidates for natural kinds, apart from a few exceptions. However, research is beginning to do so based on the types of genetic mutations found in the cancer cells.[II]

Even though subcategories like *lung cancer* may not be natural kinds, based on their cellular or physiological properties, might they be *epidemiological* kinds, based on their causal patterns of incidence, their distribution and prevalence within human populations, and so on? Since I have argued that natural kinds pertain to scientific *domains* (section 3.6), a category may be a natural kind in one domain but not another. There is a well-established causal connection between cigarette smoking and lung cancer, so it might seem as though lung cancer is at least an etiological kind based on its epidemiological properties rather than its physiological ones. The main problem with this conjecture is that on closer inspection, different subtypes of lung cancer turn out to be differentially associated with cigarette smoking; for instance, *nonsmall cell adenocarcinoma* is prevalent in both smokers and nonsmokers and may not be causally linked to cigarette smoking (though it may have a more complex causal relationship to it that is mediated by genetic or epigenetic factors). Therefore, *lung cancer* does not seem to be an epidemiological natural kind either, though there may be certain types of *lung cancer*, rather than *lung cancer* itself, that are associated with epidemiological properties, and may therefore be epidemiological natural kinds.

To summarize, *cancer cell* is a polythetic natural kind associated with one necessary property, as well as with other ("hallmark") properties that in certain combinations are sufficient for making a cell cancerous, resulting in a fuzzy kind. What makes a cell cancerous is its acquisition of further properties and its participation in a causal process, *cancer*, which is itself a natural kind of process that impedes the normal functioning of the organism.

5.6 ATTENTION DEFICIT HYPERACTIVITY DISORDER (ADHD)

In this chapter, I began by looking at physical and chemical kinds (*lithium*, *polymer*), followed by a biochemical kind (*virus*), and then two medical or physiological kinds (*cancer cell* and *cancer*). The next case to consider

---

[II] For example, one recent study has identified ten different types of breast cancer based on the mutations found in the cancer cells (Curtis *et al.* 2012).

derives broadly from the realm of medicine, but rather than pertaining directly to physiology, it is a psychiatric condition that, at least for now, does not seem to correspond neatly to a type of neural or neurophysiological condition. Psychiatry is a branch of medicine but it is also, of course, a social science, since it is concerned at least partly with understanding human beings and their interactions with each other. Hence, this category also serves as an example of a category in the social sciences, albeit one with strong connections to the biological sciences.

As is well known, the standard guide to psychiatric taxonomy is the *Diagnostic and Statistical Manual of Mental Disorders* (DSM) of the American Psychiatric Association (APA), which has been revised several times and whose fifth edition (DSM-V) is due to appear in 2013. Though the DSM has standardized the classification of mental disorders over the past six decades, it has also been substantially overhauled with each new edition, giving rise to understandable skepticism over the validity of its taxonomic categories. I will not attempt to give a history of the DSM or of its various revisions, but I will try to provide a brief overview of the history of the category of *attention deficit disorder* or its close relatives since the second edition, DSM-II, was first published in 1968. In that edition, a condition was introduced under the title "Hyperkinetic reaction of childhood (or adolescence)" and simply described as a disorder "characterized by overactivity, restlessness, distractibility, and short attention span, especially in young children; the behavior usually diminishes in adolescence" (APA 1968, 50). In the third edition (DSM-III) published in 1980, this was replaced by "Attention Deficit Disorder," which was divided into two subtypes, with hyperactivity and without, with the proviso that "it is not known whether they are two forms of a single disorder or represent two distinct disorders" (APA 1980, 41). This edition also identified the "essential features" as consisting of "developmentally inappropriate attention, impulsivity, and hyperactivity," going on to give examples of such behavior at home and at school (APA 1980, 41). The manual also listed several diagnostic criteria for each of the three "essential features" – inattention, impulsivity, and hyperactivity – and specified that three of five criteria should be satisfied for inattention (e.g., "often fails to finish things he or she starts"), three of six for impulsivity (e.g., "shifts excessively from one activity to another"), and two of five for hyperactivity (e.g., "has difficulty sitting still or fidgets excessively"). It also added that the onset must have occurred before age 7, the duration must have been at least 6 months, and the symptoms must not be due to other disorders such as schizophrenia (APA 1980, 43–44). In the revised third edition, DSM-III-R (1987), this characterization was modified significantly, with fourteen criteria listed for the disorder as a whole without further differentiation into subgroups, eight of which must have been detected for a diagnosis of ADHD (APA 1987, 52–53). More than half of these criteria remained more or less the same, including those mentioned above and others such as, "is easily distracted by extraneous stimuli" and "often does not seem to listen to what is being said to him or her." In DSM-IV (1994) and its lightly amended version DSM-IV-TR (2000), "Attention-Deficit/Hyperactivity Disorder" was retained as a condition with many of the same diagnostic criteria found in DSM-III. The most important innovation was that the diagnostic criteria were again split into groups, this time into two subsets of nine criteria each. The first was subtitled "inattention," and this subcategory is sometimes labeled "ADHD-I," while the second was subtitled "hyperactivity-impulsivity," and is sometimes labeled "ADHD-HI." The new criteria for ADHD-I included "often fails to give close attention to details or makes careless mistakes in schoolwork, work, or other activities" and "often has difficulty organizing tasks and activities". Meanwhile, ADHD-HI included such criteria as "often fidgets with hands or feet or squirms in seat" (for hyperactivity) and "often has difficulty awaiting turn" (for impulsivity). Moreover, the two subsets were explicitly introduced as disjunctive, and each carried a proviso that six or more of the nine symptoms must have been present "for at least 6 months to a degree that is maladaptive and inconsistent with developmental level" (APA 2000, 85). In addition to these two sets of disjunctive criteria, there were four additional diagnostic criteria stating, roughly, that some of the symptoms must have been present before 7 years of age, they must be present in at least two settings, there must be significant impairment in functioning, and they must not be better accounted for by another disorder. As of this writing, the DSM-V is still in preparation, but the proposed changes to the category *ADHD* seem rather limited, one of the main ones being that certain paradigmatic behaviors are appended to each criterion. For instance, the first criterion mentioned for a diagnosis of ADHD-I, involving making mistakes in schoolwork, would contain the parenthetical addendum, "e.g., overlooks or misses details, work is inaccurate."[12]

The characterizations of *ADHD* in successive editions of the DSM would seem to be highly problematic for a taxonomic category that is a candidate for being a natural kind. Even if one allows, charitably, that the

---

[12] Proposed revisions to the DSM, which are slated to appear in DSM-V, can be found at: www.dsm5. org/ProposedRevisions/Pages/proposedrevision.aspx?rid=383#.

successive overhauls represent genuine advances in understanding rather than reflecting the stipulations of successive groups of psychiatrists, other problems remain. The first problem has to do with some of the seemingly arbitrary lines that are drawn by way of characterizing it, such as the requirement that the symptoms be present before age 7, or that six or more of the nine symptoms be present, or that they be present for at least 6 months. (One is tempted to ask, why not age 6, why not five symptoms, and so on?) Another problem with the characterization, at least that found in the DSM-IV, is the very disjunctive structure of the category, which is widely considered to be an obstacle for a category's being considered a natural kind, and rightly so (as I first argued in section 1.4). A third problem concerns the presence of expressions such as "maladaptive" or "significant impairment in functioning," which are normative or evaluative expressions and seem (at best) to presuppose a set of social norms (which may be culturally variable), or (at worst) be subjectively determined by the predilections of clinicians. It would seem as though *ADHD* would be disqualified from being a natural kind on the basis of at least some of these features.

These are significant problems with the characterization of *ADHD* as it appears in successive editions of the DSM. Indeed, the substantive revisions in the understanding of *ADHD* (and its precursors) over the past four and a half decades may itself occasion skepticism about this category. But I will argue that these difficulties do not necessarily rule out *ADHD* as a psychiatric natural kind. The first point to make in this connection is that the DSM is not a theoretical treatise and that it does not purport to give a causal account of psychological conditions or processes, but is meant rather as a diagnostic tool for use by psychiatrists in a clinical setting. It bears a similar relation to psychological science as a field guide to birds does to ornithology. It is an instrument for recognizing certain psychiatric disorders or conditions by a clinician, not a scientific description of each category with attention to its explanatory and predictive import.[13] Since the emphasis is on diagnosis and treatment, one would expect to encounter a decision procedure, and diagnostic decisions may be facilitated by quantitative markers (7 years old, duration of 6 months, and so on). These seemingly arbitrary lines can be thought of not as part of the causal account of the disorder itself but as indicators that may serve to help psychiatrists and clinical psychologists to *recognize* the disorder.

---

[13] This is not to say that the DSM is atheoretical or not based on theoretical considerations, a claim which is criticized in Cooper (2004).

Nevertheless, the benchmarks used by the DSM with regard to a diagnosis of ADHD, may tip us off to the fuzziness of the kind. When there is no distinct division in nature between members of a kind and nonmembers of a kind, we may sometimes draw the line at a point that marks a salient threshold. It is apparent that like many other kinds we have encountered, in the biological as well as the natural sciences, *ADHD* and other psychiatric kinds are likely to be fuzzy kinds.

These points concerning the purpose of the DSM and the fuzziness of the kind itself may alleviate the first objection raised to the characterization of *ADHD*, but they do not clearly speak to the other two objections, having to do with the disjunctive nature of the category and its apparent normative dimension. To address these issues, it is necessary to go beyond the DSM to look at recent research on ADHD that attempts to give a theoretical account of the psychiatric condition and to understand its causal underpinnings. One prominent research program that has proposed a causal mechanism for understanding the phenomenon of ADHD posits that the core problem in ADHD is "deficits in executive functions" (Goldstein and Naglieri 2008; cf. Barkley 1997). Though the notion of "executive function" is not fully understood, it has been the subject of considerable theorizing in psychology and cognitive science over the past few decades. It is hypothesized to denote "higher-order, self-regulatory, cognitive processes that aid in the monitoring and control of thought and action," and the processes involved include "inhibitory control, planning, attentional flexibility, error correction and detection, and resistance to interference" (Carlson 2005, 595). Admittedly, there is some debate as to whether executive function names a single cognitive ability or a broad array of such abilities, or perhaps a combination of two, such as inhibitory control and working memory (Carlson 2005, 596). Still, this construct has been "operationalized" to some extent, since there are a number of cognitive tasks that are quite widely accepted as tests for executive function, such as anti-imitation tests in which participants first undertake trials in which they imitate an experimenter's actions but then must continue by performing the opposite action. This and related tasks require participants to inhibit their initial response and to carefully control their impulses by thinking them through and planning ahead.

Whether or not *executive function* itself turns out to be a single psychological kind, this operational understanding of *executive function* has been used by some researchers as a basis for a theoretical account of ADHD, which understands it precisely as a deficit in the capacity of executive function.[14] The hypothesized deficit in executive function (P_1)

is associated by Goldstein and Naglieri (2008) with five properties, all of which are thought to characterize ADHD patients. First, *impulsivity and lack of planning* (P_2): Individuals with ADHD have difficulty in inhibiting action while thinking through the best way to achieve their goals and in using working memory to construct a plan and monitor its execution. Second, *inattention* (P_3): These individuals also struggle to sustain their efforts and to invest in tasks that must be completed. Third, *hyperactivity* (P_4): They are restless and do not inhibit fidgety behavior. Fourth, *problems with modulating gratification* (P_5): These individuals are driven towards immediate, frequent, and predictable consequences, require brief repeated payoffs, and have difficulty working toward a long-term goal. Fifth, *problems with emotional modulation* (P_6): They are also emotionally aroused more quickly than others, whether positively or negatively, indicating lack of control over their emotional reactions. According to this research program, these cognitive and behavioral features of individuals with ADHD are directly causally linked to deficits in executive function. It is not clear whether the first property is causally responsible for all the rest or whether there are more complex causal relations among these properties, but these questions are subjects of ongoing research.

This research program, which regards ADHD as involving deficits in executive function, also considers that the disjunctive characterization of the condition in the DSM is mistaken. Goldstein and Naglieri (2008) propose that ADHD-HI involves a different deficit than ADHD-I and ought to be categorized differently. The reason is that "children with ADHD-HI are characterized as having a failure of behavioral control, whereas children with ADHD-I are described as having a failure of selective attention" (2008, 865). Moreover, the inattentiveness involved in these cases is different from that identified with P_3, since they argue that the failure is one of behavioral control rather than attention proper. Individuals with ADHD-HI are capable of attending to stimuli that interest them but not to ones that do not; "they can attend to their favorite computer game but have considerable problems staying on task in the classroom" (Goldstein and Naglieri 2008, 865). On the basis of these considerations, Goldstein and Naglieri (2008, 865) suggest not only that

---

[14] Here, as elsewhere, in considering whether *ADHD* might be a natural kind, we need to bracket questions as to whether certain related categories (e.g., *executive function*) pick out natural kinds. But, in this case, the assumption is rather more controversial than in some of the other cases discussed (for example, when we implicitly assumed in discussing the category *cancer cell* that *gene* was a natural kind in its own right).

the disorder be split into two separate disorders, but also that the one involving executive function be renamed to reflect its different nature:

> It would be more logical and consistent with the symptomology to describe children with ADHD-I as attention deficit because they do have problems with selective attention, but those with ADHD-HI require a different label – perhaps not having an attention deficit but rather a self-regulation deficit.

In fact, there seems to be broad agreement among researchers attempting to provide a theoretical understanding of ADHD that the disjunctive characterization of ADHD is incorrect and that there are two distinct disorders at issue. Even researchers who posit a different theoretical account of ADHD and attribute it to a distinct psychological mechanism agree on this point.[15] Many of the researchers interested in understanding the causal basis for the disorder are evidently not content with the disjunctive nature of the disorder as described in the DSM, positing instead two distinct disorders, each with a set of characteristic properties, different enough to warrant separate labels.

The third problem that I mentioned has to do with the apparent normative or evaluative dimension of the category, as evidenced by the diagnostic criteria in the DSM. When one looks at the causal or theoretical account of ADHD sketched above, by contrast, these normative elements do not stand out, at least not overtly. The properties listed above (P_1–P_6) do not make explicit reference to norms of behavior, though they do incorporate comparative judgments (e.g., "emotionally aroused more quickly"), which may be taken to pertain to covert norms or normal behaviors. It is tempting to say that the norms to which ADHD individuals are being compared can be understood to refer to a statistical average or median, as opposed to a relative or subjective judgment as to what constitutes "normal" behavior. These types of comparisons are prevalent in other areas of science, so the comparative element in accounts of *ADHD* (and other psychiatric categories) need not pose a special problem. But the obvious reply to this claim is that even identifying emotional arousal or restlessness as a relevant difference between human beings involves an evaluative judgment. Why not ignore such differences in children, as we

---

[15] A prominent rival theory of ADHD proposes that the disorder is caused by a hypofunctioning dopamine system and attempts to describe it in more purely behavioral, as opposed to cognitive, terms. Nevertheless, Sagvolden *et al.* (2005, 402) concur that "symptoms and developmental course indicate that the present ADHD diagnosis consists of two separate disorders probably with separate etiology." They also focus on explaining ADHD-HI and propose renaming it "reinforcement-extinction disorder" (RED).

usually ignore differences in their favorite colors or in their ability to wiggle their ears? If one says that the former differences are disruptive to those around them or indeed are harmful to their own success, does that not reflect our own parochial cultural norms and the specific social institutions that we have established? It is true that cognitive and behavioral traits can be detrimental to a child's success and flourishing in the context of some types of social arrangements and not in others. Since these social arrange-ments are socially and culturally variable, a trait that may impede a child's success in one social environment may not in another.[16] Indeed, the salience of that trait may itself be a function of the social backdrop; some human communities may not even recognize differences in emotional arousal or restlessness as salient psychological features. But within other social contexts, these traits have real causes and effects that make a difference to the lives of individuals and those around them. This is not just a reflection of the subjective preferences of psychiatrists, clinical psychologists, or social workers. Social and human kinds are relative to human societies and cultural environments, but their causal powers are real and objective relative to those environments. The causal patterns in which they participate are partly a matter of the relation of individuals to their surroundings, which is to say that they are a relational or extrinsic matter. But they are nonetheless real, just as some biological phenotypic features are adaptive in some environments and neutral or maladaptive in others.

In some respects, the norms relative to which psychological kinds are identified are analogous to those relative to which biological kinds are delimited, though defining the abnormal may be more problematic in a psychological and social setting than it is in a biological or physiological setting, since death has a convenient finality when it comes to physiology (cf. Murphy 2006, 85–87). But in the case of many social kinds, there is often an added *moral* dimension to them that puts pressure on us to modify the category in certain ways. As I argued in section 4.7, given the moral censure associated with the category *child abuse*, there is an incentive for some groups to classify behaviors that they object to as *child abuse*, in order to condemn those behaviors morally (and sanction them legally). When it comes to *ADHD-HI*, the principal values associated with the category do not seem to be moral but rather *pragmatic*, in the sense that ADHD-HI individuals may be thought to be less capable of acting in their

---

[16] The cultural dimension is acknowledged by many researchers in this area: "Familial, social, academic, and vocational demands *of a fast paced culture* require a consistent, predictable, independent, and efficient approach to life" (Goldstein and Naglieri 2008, 863; emphasis added).

own self-interest or succeeding in their own endeavors, rather than being held to be morally deficient in some ways. (Indeed, treating the condition as a psychiatric disorder may have the effect of taking it out of the ethical sphere and ceasing to regard some of its symptoms as moral failings – e.g., deficits in modulating gratification.)

Although moral norms may not put pressure on changing the boundaries of this particular category, there may be other social factors that do so. Identifying the disorder as a distinct condition with negative consequences for the child (in some social settings) will tend to cause many parents and caregivers to seek treatment for their children's condition, putting a premium on treatments or cures. This, in turn, may give rise to pressures to alter the boundaries of the category to serve certain interests. This point can be illustrated by some critiques of ADHD that regard it to be an artificial category, one that serves the interests of the pharmaceutical corporations. Some researchers have alleged that far from being a distinct condition or disorder, the symptoms attributed to ADHD are, in fact, symptoms of a variety of different underlying "medical, emotional, and psychosocial conditions affecting children" (Furman 2008, 775). This criticism goes on to suggest that the reason that ADHD has been considered an "identifiable disease" with a single neuropsychological basis is that this claim serves the financial interests of pharmaceutical firms. If ADHD is believed to have a single neuropsychological basis, then it will be thought to be treatable with a single chemical substance, which can be patented and marketed to patients. Thus, corporations that have an interest in promoting specific drugs subtly manipulate research findings to serve their financial interests. Furman (2008, 780) claims: "Pharmaceutical firms that manufacture ADHD medications play a large role in promotion of ADHD as a disease requiring medication." She thinks that this manipulation takes place because many scientists involved in ADHD research have financial ties to the pharmaceutical corporations, receiving funds for speaking, consulting, or conducting research. Since the researchers are beholden to the corporations, their financial ties tend to produce a "self-interest bias". Although it is not deliberate, the bias is pervasive and influences the way in which these researchers process and interpret information. Furman (2008, 780) concludes:

Physicians should consider the possibility that extensive pharmaceutical funding has essentially controlled research and treatment for children with symptoms of hyperactivity and inattention and has driven an explosion in the number of children who must be diagnosed as ADHD.

Furman thinks that researchers who have ties to pharmaceutical corporations that manufacture ADHD medications (according to her, thirteen of the twenty-one individuals who created the DSM-IV criteria for diagnosis of ADHD) have an incentive to interpret the empirical evidence as suggesting that ADHD is a single disorder that is treatable by a single drug.

Furman (2008) may be right that the claim that *ADHD* is a single natural kind could be affected by the financial ties of some researchers to pharmaceutical corporations. It may be that many researchers have been influenced by prescription drug manufacturers to identify a single disorder where there are many, or to draw the boundaries of the category in a certain way. These biases are not unknown outside psychiatry and the social sciences, though they may be more prevalent in the social sciences due to the fact that human interests tend to be more prominent when human or social categories are involved, and hence the stakes are higher for human beings.[17] But when a research agenda is being driven by nonepistemic purposes, such as financial interests, these biases are also capable of being uncovered and corrected, as I argued in section 4.7. Social scientists are capable of identifying biases in the formulation of a taxonomic category, in roughly the way that other empirical inquirers do – namely, by determining whether the category plays a genuine explanatory role and has predictive value. To the extent that it does, the category will have been vindicated. Since at least some of the recent research suggests that the category *ADHD* corresponds to two distinct natural kinds (which are not subkinds of a single natural kind), these researchers clearly reject the claim that it is a single disorder. This already provides some evidence that such influences are not inevitable and are capable of being overcome. Nonetheless, we are forced to conclude here, as before, that our opinions as to the existence of a natural kind and its boundaries are capable of being influenced by nonepistemic interests, especially in the social sciences. When this occurs, scientific categories will not correspond to natural kinds, and though such biases may persist, they are capable of being corrected in the course of scientific inquiry.

In this connection, it is worth stressing that if it turns out that *ADHD-HI* is indeed a distinct psychiatric natural kind, as some of the recent research suggests, it may not correspond to a single *neural* natural kind. Researchers who make the former claim do not always make the

---

[17] For a classic discussion of how "contextual values" can influence scientific conclusions in a variety of scientific fields, see Longino (1983).

latter.[18] That is simply because a psychological or cognitive condition, such as a deficit in executive function, may be multiply realizable with respect to neural states or processes. This is analogous to the claim that a single fluid mechanical property like *viscosity* is multiply realizable with respect to molecular structure (as I argued in Chapter 3). There may well be different neurochemical or neurophysiological reasons for why individuals have deficits in executive function, just as there are different reasons for why liquids and gases are Newtonian fluids. Yet these differences may not be germane when assessing the cognitive and behavioral profiles of those individuals. Different neural processes may give rise to similar psychological effects (especially across different contexts). This is not to deny that these differences may matter for certain purposes. For instance, some of these neural processes might be capable of being modified by the action of a specific chemical, but others may not be. Would that not show that there is no single natural kind in question? Not necessarily, for reasons already encountered in previous chapters. There may yet be important commonalities in the causal processes that these individuals participate in from a psychological or psychiatric perspective, and the category *ADHD* may enable us to devise unified explanations or make important predictions concerning the cognitive and behavioral development of these individuals, regardless of their neurochemistry, just as there are higher-level commonalities in cases of multiple realizability in other domains. These commonalities could also lead us to cognitive and behavioral treatments of the condition that can be used to alleviate the condition or affect its developmental trajectory.

There is an additional complication to this picture of causation and causal processes in psychiatry. In many psychiatric conditions, the neurobiological causes interact with the cognitive and behavioral ones in surprising ways. The causal properties associated with a psychological condition or disorder may also be entangled in certain ways, sometimes reinforcing or undermining each other. They may not be related merely as effects of a common cause or sequentially in a hierarchy of causes, but may also involve causal feedback loops. In this case, as a result of problems modulating gratification, some ADHD-HI individuals tend to have negative social interactions with others, which then cause these individuals to

---

[18] But some researchers do. Faraone (2005, 5) argues, based on a review of twenty-eight neuroimaging studies of ADHD children, that the condition involves dysfunction in fronto-subcortical pathways: "The most common findings are smaller volumes in the frontal cortex, cerebellum, and subcortical structures, especially striatum."

avoid the aversive consequences of these interactions, leading them to be more sensitive to negative reinforcement than to positive rewards, and further exacerbating their inability to modulate gratification (Goldstein and Naglieri 2008, 863).[19] In addition to multiple realizability, there are feedback loops among causes within the same domain and interactivity among domains. Indeed, there may even be a crosscutting relationship among domains, which would lead to a many–many relationship among psychological and neural conditions (rather than a many–one relationship).

The current research programs that consider ADHD, or rather ADHD-HI, to be a single disorder may well turn out to be incorrect. My aim in this section has been to suggest, first, that there is at least some evidence to suggest that *ADHD-HI* is a psychiatric natural kind, and, second, that the standards and principles for identifying natural kinds in a social science like psychiatry are similar in their general features to those deployed in the other sciences that have been examined. Although treatment is a central concern in psychiatry, researchers involved in discovering psychiatric categories are interested in understanding the causal relationships among psychological properties, which enable them to understand why properties cluster as they do. Natural kinds play a central role in causal networks. They are associated with sets of properties that are causally linked in reliable ways.

5.7 CONCLUSION

In this chapter, I have considered several candidates for natural kinds in the natural, biological, and social sciences. In all cases, the current state of scientific research suggests that the categories in question are indeed natural kinds, though we have also encountered some categories that do not appear to be natural kinds, including superordinate categories (e.g., *ADHD*, which includes *ADHD-I* and *ADHD-HI*) and subordinate categories (e.g., *lung cancer*). I have been relying on current scientific theories of the phenomena that I have studied, and in some cases there is no settled consensus concerning the domains involved. If the scientific theories concerned are revised radically, then some of my conclusions may be jeopardized. But, as I mentioned in section 2.2, given the corrigibility of

---

[19] Kendler, Zachar and Craver (2011, 1147) also argue for causal interaction among the properties associated with a single psychiatric kind, including the following example of a causal feedback loop: "Phobias lead to avoidance, which prevents habituation to the feared stimulus."

scientific theories, there can be no definitive answer concerning the natural kind status of any given category until the end of inquiry, when a final theory emerges. Nevertheless, what we should be concerned with is not so much the particular answers given concerning the natural kindhood or otherwise of the categories considered. What is important for this inquiry is the meta-scientific practice of taxonomy and classification; that is, the means whereby scientists go about identifying natural kinds and the characteristics they expect them to have. Closer attention to taxonomic practices in the sciences bears out some of the conclusions reached in earlier chapters. Natural kinds of entities are nodes in causal networks, serving either as starting points of branching networks or as endpoints (etiological kinds), or both. Most kinds are closely associated with one or a few central necessary properties, but few are associated with them in such a way that these properties are both necessary and sufficient for the kind in question. When instantiated or co-instantiated, these properties then give rise to or cause the instantiation of a series of others. The kind itself is typically associated with the primary properties, though it is also sometimes identified with the (often loose) cluster of properties in the entire network. What determines membership in a kind is not the possession of a requisite number of properties in the cluster but rather involvement in many of the same causal relations. These causal relations often unfold in a temporal causal process, which is set in motion by the primary properties associated with the natural kind. The causal links in these processes are often not deterministic but probabilistic. Some of these kinds are what philosophers call "functional kinds" while others are "structural kinds," though the distinction does not seem to be a deep one and is challenged by such kinds as *polymers*, which are not obviously examples of either. In some cases, there can also be interaction between the properties of a kind, which leads to feedback loops, sometimes also keeping some of the properties in homeostasis or equilibrium (as in Boyd's account of natural kinds). However, this is a characteristic of a minority of the kinds that we have encountered and is more attested to in the biological and social sciences than in other sciences.

CHAPTER 6

Kinds naturalized

6.1 NATURALISM ABOUT KINDS

Naturalism means different things to different philosophers, but one common thread is surely the injunction to take heed of the evidence from the sciences in formulating and assessing philosophical theories. If we take such evidence seriously, I have argued, it steers us away from an essentialist position concerning natural kinds and leads us towards something like a "simple causal theory" (Craver 2009) of natural kinds. On this theory, natural kinds are associated with a set of properties that are causally linked to other properties in a sequence or network. When the properties associated with a natural kind are instantiated or co-instantiated, they lead reliably to the instantiation of a number of other properties. Since they are implicated in repeatable patterns of properties, they enable us to explain and predict phenomena in the natural and social worlds.

What does it mean to take scientific evidence seriously? In line with naturalist principles, I have sampled a range of categories from a variety of scientific disciplines in order to determine what features they have in common. But this method is open to the objection that this is purely a descriptive exercise and one that makes philosophers into passive observers of the deliverances of the sciences. Worse still, unless one has antecedent standards for dismissing some specimens as illegitimate, the mindless accumulation of categories from the sciences has no prescriptive force. If one does have such standards, then these are presumably not, on pain of circularity, simply derived from the categories that one has examined from the various sciences. Hence, there must be a priori or conceptual principles that philosophers apply in determining whether a category is valid and has a legitimate claim to denoting a natural kind. This line of reasoning is supposed to demonstrate the poverty of the naturalist approach.

A full response to this objection to philosophical naturalism would require a more substantive treatment and would necessitate engagement with the considerable recent debate on this topic.[1] But I will try to address the objection in the specific context of this inquiry into natural kinds. In Chapter 1, I mentioned that I would be guided by the method of reflective equilibrium famously enunciated by Goodman (1954/1979), in which one shuttles between rules and examples, or in this case, between philosophical principles and scientific case studies, making mutual adjustments, until one achieves agreement between the two. But where do these philosophical principles come from and what is the source of their authority? There are at least two sources for the philosophical principles that guide the inquiry into the nature of natural kinds. The first source, which is of lesser importance, involves giving some consideration to past usage. In this case, that means the analysis of the concepts of previous generations of philosophers. Since the concept of *natural kind* is not a vernacular one, attention needs to be paid to the way in which other philosophers have deployed the concept. This is not a case of sheer deference to authority but a matter of recognizing that if one departs radically from previous philosophical usage, then one can be accused of changing the subject. But have I not departed radically from the doctrines of philosophical essentialism? Yes, but essentialism is by no means the only philosophical theory of natural kinds. When one looks at the way in which the progenitors of this concept used it, particularly Mill, one finds that they have very little in common with contemporary essentialists. Among more contemporary theorists who have written on the topic, I have found much to agree with in the views of many of them, though I have also registered some disagreements with most contemporary philosophers who have written on natural kinds. One should pay some attention to how the concept has been deployed by other philosophers; otherwise, one might rightly be said to be talking about something entirely different.

Another, and more important, source of philosophical principles to be injected into reflective equilibrium when characterizing natural kinds has to do with the work that we want the notion of *natural kind* to do for us. In other words, we need to ask how the concept of *natural kind* can help us achieve our philosophical and scientific goals. Here, the context that is most pertinent to the identification of natural kinds is that of scientific inquiry, since scientific disciplines aim to locate projectible categories that enable them to predict and explain the phenomena of interest. I have connected this epistemic aim to the metaphysical goal of identifying genuine patterns and configurations in nature, principally real causal

---

[1] For a defense of the "meliorative" or normative dimension of the naturalist project, see Kitcher (1992).

patterns. Therefore, one important principle that guides our investigation into the characteristics of natural kinds concerns their projectibility, and hence their centrality to causal patterns or causal networks. This concern is closely related to the methodological concerns of at least some working scientists, and it coincides quite closely with what some of them mean when they discuss "construct validity."[2]

In case the content of this second principle seems rather abstract and nebulous, an illustration of the method at work might help. At an early point in this inquiry into natural kinds, I considered the question of whether a natural kind could be structured disjunctively, and I argued that disjunctive categories do not correspond to natural kinds (see section 1.4). The main consideration I put forward for thinking so was that disjunctive categories are not projectible. It is not that philosophers have privileged access to the a priori or conceptual truth that natural kinds cannot be disjunctively structured. Rather, given that the whole point of discovering natural kinds is to discover sets of co-instantiated properties that are inductively linked to a series of other properties, disjunctive categories do not enable us to achieve this goal. Disjunction undermines project-ibility, so it defeats the purpose of having natural kinds in the first place. Moreover, this conclusion is vindicated by the fact that disjunctive kinds are rarely if ever attested in actual scientific practice, and where they are, they appear to be challenged by scientists for sound reasons.[3]

A third source of philosophical principles that constrains the account of natural kinds has to do with consistency with conclusions reached in other areas of philosophy. In particular, given the central place of causation in the simple causal theory of natural kinds, it is necessary to determine whether this theory of natural kinds is compatible with philosophical analyses of the causal relation. Some philosophers have argued that basic metaphysical assumptions about causation do not allow for causal relations to coexist in different domains of the universe or at different levels of reality. However, the existence of natural kinds in different domains, each with its own causal networks is central to this account of natural kinds. Fortunately for this inquiry, other philosophers have argued for the compatibility of causal relations at different levels or in different domains

---

[2] The classic discussion in Cronbach and Meehl (1955) is particularly apposite since they argue that valid constructs are situated in "nomological networks."

[3] As we saw in our discussion of *ADHD* (in section 5.6), at least some researchers who attempt to achieve a theoretical and causal understanding of the condition have cast doubt on the disjunctive nature of the category as it is described in the DSM-IV and treat each disjunct as corresponding to a distinct natural kind with different properties.

on independent grounds. As I indicated in section 3.3, this is true, for example, of the interventionist theory of causation, as well as some others. Therefore, there is no blatant contradiction here with an uncontroversial tenet derived from a different area of philosophical theorizing.

Based on this methodological approach, I have defended a philosophical account of natural kinds that has close affinities to some of the theories proposed by other philosophers, though I think it also contains some novel elements. But there are some remaining issues that still need to be addressed in this chapter. In the following sections, I will consider further the precise relationship between natural kinds and properties, and then I will try to elaborate on the connection between natural kinds and causality. In addition, I will address the question as to whether this naturalist account of natural kinds is a realist one, and I will also justify further the claim (made in Chapter 4) that mind-independence is not criterial for realism.

6.2 PROPERTIES AND KINDS

We started with a very widespread philosophical conception of natural kinds according to which each natural kind K is associated with a set of (projectible) properties, $\{P_1, \ldots, P_n\}$. But this simple picture has been found wanting mainly because it leaves out the relationships that exist between the properties associated with a kind. Crucially, in all the cases that we have encountered, there is a causal link between properties, with one or a few of the properties being causally prior to the others. What characterizes natural kinds is that, even when one or a few properties are central to a kind, there are a number of other properties associated with that kind that are causally related to them. It is this network of properties that seems to distinguish natural kinds from nonnatural kinds. The causal relations between the properties in the network ensure that natural kinds are projectible and play a central role in inductive inference, as I have elaborated in previous chapters. In short, the important epistemic role of natural kinds is underwritten by their metaphysical status.

In all the cases of natural kinds that we have examined, some properties associated with the kind have causal priority over others. In section 2.6, I distinguished what may be called the "primary" causal properties of natural kinds, which are causally prior, from what may be called the "secondary" properties that follow from them.[4] This implies that there

---

[4] Needless to say, "secondary properties or qualities" in this sense are not the same as the secondary properties posited by Locke (1689).

are two different ways in which properties are co-instantiated when it comes to natural kinds. The primary properties are co-instantiated with the secondary properties because they are related as cause and effect, but the primary properties themselves are co-instantiated not because they are directly causally linked. For instance, in different isotopes of chemical elements, atomic number and mass number are not related as cause and effect (though they are not completely independent of one another). Is there a general account of how the causally prior properties associated with natural kinds come to be co-instantiated? In section 4.4, I considered an answer inspired by Millikan – namely, that some do so as a result of natural law while others do so as a result of a copying process. But I argued that this distinction does not run very deep and does not mark off two disjoint types of natural kind. I concluded there that the main difference, such as it is, between eternal kinds and copied kinds is that members of the former arise independently of other members of their kind, while members of the latter do so partly as a result of interaction with other members of their kind. Hence, it seems that all that we can say is that some combinations of properties in the universe are allowable and others are not, and that is ultimately a matter of natural law. Moreover, some of these combinations lead to the instantiation of a range of other properties. These combinations of properties are the ones associated with natural kinds.

Throughout this book, I have spoken in terms of natural kinds being "associated with" a set of properties, but this discussion may now suggest that natural kinds should in fact be identified with a set of properties – namely, the primary properties, or those that have causal priority over the others. What is the precise relationship between natural kinds and the primary properties with which they are associated? Are the primary properties identical with the natural kind itself? In a few of the case studies I have examined, there was one property that was central to the kind from which all other properties derived causally, whether directly or indirectly. But in at least some of those cases, that property was not sufficient for the instantiation of the other properties associated with the kind, and the relationship between the kind and its associated properties turned out to be more complex than might seem at first. Many natural kinds are cluster kinds; in other words, they are polythetic rather than monothetic. In fact, I argued that this is even true of perhaps the least controversial exemplars of natural kinds, the chemical elements. In general, the relationship between the natural kind and its primary properties is not one of necessity and sufficiency.

There are several further complications that need to be mentioned in attempting to characterize the relationship between natural kinds and their primary properties. One recurring feature of the natural kinds that we have encountered is that the properties associated with them are typically determinates rather than determinables (e.g., not *atomic number*, but *atomic number 3*, for the natural kind *lithium*). Generally speaking, determinable properties, such as *atomic number, mass*, or *viscosity*, are not associated with natural kinds. Clearly, not all determinate properties, such as having a *melting point of 180.5 °C*, are natural kinds, simply because they do not, by themselves, lead to the instantiation of many other properties. A second feature of some of the primary properties associated with natural kinds is that they are complex functions of other properties or elaborate logical constructions out of simpler properties rather than simple properties in their own right (e.g., the principal property associated with the natural kind *Newtonian fluid* is that *viscosity remains constant with a change in the applied force*). Third, even in some of the basic physical and chemical sciences, the properties associated with a natural kind issue in fuzzy boundaries (and hence in fuzzy kinds). This is true of the properties associated with some chemical compounds, since chemical isomers like *ethanol* and *dimethyl ether* (compounds with the same elements in the same proportions but different structures) shade into one another depending on continuous variables like internuclear distances and angles between bonds (Hendry 2006), as mentioned in section 2.3. It is also true of the chemical kind *polymer*, which has vagueness built into its necessary and sufficient property (*compound consisting of large macromolecules with repeating units*); there is no determinate answer as to how large the molecule of a polymer has to be or how many repetitions it has to have to qualify as a polymeric compound. Finally, in some cases, some of the primary properties associated with a natural kind require a certain context for the instantiation of the other causal properties characteristic of the kind, making the properties in question dispositional rather than categorical. Though many dispositional properties are bona fide properties (indeed, some philosophers would claim that all properties are dispositional or at least partly dispositional, as mentioned in section 1.2), the dispositional properties in the case of some natural kinds are very specific to a certain context. For the kind *virus*, one of the necessary properties is *having a genome capable of making mRNA that is readable by the ribosome of a host cell*. This idiosyncratic necessary property is narrowly linked to a specific function and indexed to other kinds with which virus particles interact (*mRNA, ribosome*, etc.). These complications combined suggest that the relationship between natural kinds and their associated properties cannot be considered one of straightforward identity since there is a certain looseness of fit between natural kinds and their associated properties; furthermore, the class of natural kinds is distinct from the class of real properties.

There are two tiers of properties associated with natural kinds, the primary properties, which are co-instantiated because this is allowable according to natural law, and the secondary properties, which follow causally from them. The relationship between the primary and secondary properties sometimes unfolds in a regular and repeatable temporal sequence. In many of these cases, a natural kind is closely associated with a distinctive causal process. I have argued that these processes can be considered natural kinds in their own right. Natural kinds of causal process, which some other philosophers have also endorsed, involve the instantiation of properties in a regular temporal sequence (e.g., *cancer*). But in these cases, it is sometimes difficult to draw a line between the primary properties and the secondary properties of the natural kind of individual or object associated with the process, as we found in the case of the natural kind *cancer cell*, since there was no clear demarcation between cancerous and noncancerous cells. This is especially so because, in most cases, the primary properties associated with a natural kind, which have causal priority over the others, do not strictly causally *determine* the other properties associated with that kind, but are linked to them by causal relationships that are probabilistic. In addition, those properties themselves often do not determine the ones further down the causal chain. Sometimes the relationship between the primary and secondary properties associated with a natural kind is a cyclical one. An instantiation of the primary properties can lead to the instantiation of the secondary ones, which in turn leads to another instantiation of the primary properties. This kind of cycle is perhaps most prevalent in biology, and this pattern is manifest in such kinds as *virus* and *larva*.

Associated with each natural kind there is often a network of causal properties arranged in a hierarchy of causal priority, in which the causal relationships are nonstrict and are subtly contingent on contextual factors. This set of associated properties can be described in retrospect as a cluster or a weighted cluster of properties, with weights reflecting the strength of the causal links. But that characterization leaves out something important about the causal structure of the kind. Rather than a simple set or list of properties, possibly with weights attached, the structure of a natural kind is more adequately captured by a network in which these causal relationships are explicitly indicated. Therefore, natural kinds are better understood as nodes in causal networks, rather than mere clusters of properties, even weighted clusters.

6.3 CAUSALITY AND KINDS

In the previous section, I argued that a natural kind is associated with a set of properties that, when co-instantiated, lead causally to the instantiation of a series of other properties, sometimes as part of a repeatable temporal process. It has become increasingly apparent that causation plays a major role in the individuation of natural kinds, and that their causal structure is what is distinctive about them. In this section, I will examine further the relationship between natural kinds and causality, asking first why causality is so important to something's being a natural kind, and also whether it is necessary for a kind being natural. Then, I will reconsider the question of whether there are degrees of naturalness among kinds in light of the causal structure of natural kinds.

For this account of natural kinds to succeed, there must be causal relations not just in the smallest microphysical domain but in other domains as well. Otherwise, there cannot be natural kinds in those domains. This goes against the conclusion of Kim's "causal exclusion argument," which I resisted in Chapter 3. Apart from some suggestions made in discussing the "causal exclusion argument," I have not proposed or endorsed an account of the causal relation. It may well be, as Kim claims, that an account of causation in terms of "generation" or "production" accords naturally with that argument and perhaps even forces us to accept the conclusion of the causal exclusion argument. But other philosophers have argued that there are alternative accounts of causation that do not force such a conclusion. For example, proponents of the "interventionist" account of causation have argued that that account allows for causation at different levels, and some versions may even lead to the exclusion of the microphysical level in favor of the macrophysical (see section 3.3). Since the attempt to justify a theory of causation is beyond the scope of this book, I will continue to assume that an account of causation is available that would allow for causal relations in domains beyond the smallest microphysical domain.

The content of the claim that natural kinds are individuated in terms of causal relations can perhaps be better appreciated by contrasting it with a couple of notable alternatives – namely, that kinds can be held together by virtue of logical connections or conventional links. In section 2.2, I followed Mill in accepting that a natural kind cannot be associated with a set of properties all of which simply follow deductively from one or a small set of other properties. Then, in section 4.6, I argued that the category *permanent resident* did not correspond to a natural kind on the grounds that the properties associated with the kind are conventionally associated with it as a matter of legal statute. In both of these cases, if properties are associated with a kind on the basis of logical or conventional links, we do not have the impression that this marks a real grouping of things in the world. In the first case, if some properties follow deductively from others, then they are in an important sense already present in those others; they do not constitute anything new or convey novel information about reality. In the second case, we have associated one set of properties with another by fiat; we have not discovered these properties in the world. Therefore, these types of relations do not track important aspects of reality. But on a more positive note, why do *causal links* qualify as indicating a real grouping of things in the world? It is difficult to provide a fully satisfactory answer to such a basic question, but two considerations may help to dispel some of the doubt. If science is the enterprise devoted to gaining knowledge of the universe, and if science is primarily engaged in discovering causal relationships, then this should lead us to think that causal relations are privileged features of reality. Meanwhile, it is widely accepted among philosophers that something is real if (and sometimes only if) it is capable of making a difference to causal interactions or causal processes. What has been called the "causal criterion of reality" (Kistler 2002; cf. Armstrong 1997) can be traced back to Plato's *Sophist* (247d–e). In that dialogue, the Eleatic Stranger puts forward the following criterion for "real existence":

My notion would be, that anything which possesses any sort of power to affect another, or to be affected by another, if only for a single moment, however trifling the cause and however slight the effect, has real existence; and I hold that the definition of being is simply power.[5]

The same basic principle has also been advocated by Kim (1998, 119), who phrases it as follows: "a plausible criterion for distinguishing what is real from what is not real is the possession of causal power." To be sure, this principle has its critics (e.g., Cargile 2003), and it may not be a good idea to dismiss anything that does not have causal efficacy as being unreal. Realists about numbers, aesthetic values, moral principles, and propositions may insist that these things are real notwithstanding their lack of causal powers. But if the point of identifying natural kinds is to discern real divisions in nature, causality is commonly acknowledged to be, in Hume's memorable phrase, the "cement of the universe." That is why it is

---

[5] This quotation is taken from the Jowett translation, available online at: www.gutenberg.org/files/1735/1735-h/1735-h.htm.

appropriate to associate natural kinds with networks of causal properties and to interpret real patterns in terms of real *causal* patterns.

There is a different kind of challenge to the idea that causal relations are central to natural kinds, which is supposed to be inspired by science itself. The claim has often been made that causation is a notion that is gradually disappearing from the mature sciences, specifically contemporary physics, and that fundamental physical theories do not employ causal notions. This claim was perhaps made most famously by Russell (1913, 1), who wrote categorically that "the reason why physics has ceased to look for causes is that in fact there are no such things." A few decades later, he softened his stance on this question somewhat, though he did not abandon it completely. In a later work, Russell (1948, 453) noted: "The concept 'cause,' as it occurs in the works of most philosophers, is one which is apparently not used in any advanced science," but he went on to allow that the "primitive concept" of causation "still has importance as the source of approximate generalizations and pre-scientific inductions." More recently, this position has been given an interesting twist by Ladyman and Ross (2007, 286), who make a critical distinction between fundamental physics and the special sciences:

Fundamental physics is in the business of describing the structural properties of the whole universe. These properties are not causal relations. Special sciences and fundamental physics are thus mainly different from each other in a way we find it very natural to express by saying: fundamental physics aims at laws, whereas special sciences identify causal factors.

This strict separation between fundamental physics and the special sciences is motivated by their stance of structural realism towards the field theories of current physics. Unlike many philosophers, Ladyman and Ross are not reductionists who deny the existence of properties, kinds, and causal relations in the domains of the special sciences. Quite the contrary, they subscribe to what they call "Rainforest Realism," according to which the special sciences track real patterns (in Dennett's sense). Nevertheless, they think that the patterns captured by fundamental physics are crucially different from those ascertained by the special sciences, in that the former are not causal but structural.

This position poses a challenge to the simple causal theory of natural kinds. Although I have not paid close attention to elementary particle physics, I have been arguing that causation is a common thread that runs through the domains of all the sciences, and that it is central to the natural kinds identified in each of these domains. If that does not apply to fundamental physics, this would constitute a challenge to the account of natural kinds that I have been defending. How should we respond to this challenge by Ladyman and Ross? First, it would be remiss not to point out that the research programs of contemporary high-energy or elementary particle physics are still very much in flux – some would even say in crisis (Smolin 2006). The point here is not just the familiar one that all scientific theories are corrigible but also that the paucity of evidence renders much of contemporary fundamental physics highly speculative. It is therefore premature to conclude that fundamental physics is not concerned with discovering causal relations. However, it is undeniable that quantum physics has substantially altered our understanding of causality and has led to the recognition of the existence of at least some events that are strictly uncaused, as well as the possibility of backwards causation or superluminal causation. Still, this does not constitute a wholesale abandonment of the concept of causation, and many physicists and philosophers of physics continue to regard causality as being importantly involved in the realm of elementary particles (see, e.g., Smolin 2006, 238–247). But even if one grants for the sake of argument Ladyman and Ross' interpretation of fundamental physics, it does not force us to abandon the simple causal theory of natural kinds. It seems that there are two possible conclusions that could be drawn. The first would be to say that the categories of elementary particle physics correspond to natural kinds and that the rest do not (and that causality is not therefore a feature of natural kinds at all). As I have already indicated, this is not the way that Ladyman and Ross themselves go. But other philosophers, of a microphysical fundamentalist persuasion, may take their account of the distinction between fundamental physics and the special sciences to be a reason for concluding that the causal patterns tracked by the special sciences are mere illusions and that the relationships of microphysics are privileged precisely because they are not causal but structural. There would be a certain irony in this since microphysical fundamentalists like Kim tend to privilege the domain of the ultimate constituents of the universe because they take causation to be exclusively manifested at that level. Another way to accommodate this conclusion would be to agree with Ladyman and Ross that there are real patterns in these different domains but that the real patterns in fundamental physics are structural, whereas those in the domains of the special sciences are causal. One might then posit two classes of natural kinds that constitute real features of the universe for different reasons or on different grounds. (It is not clear whether this is precisely the way that Ladyman and Ross themselves would put it.) There is no way to rule this out decisively until we have a definitive theory of fundamental physics and a consensus on its proper interpretation. But even if there is a divide between fundamental physics and the special sciences when it comes to the role of causality, the simple causal theory of natural kinds would still apply to all but the smallest microphysical domain.

Before leaving the topic of causality, I will try to relate it to a question that I have touched on before – namely, that concerning degrees of naturalness among kinds. On a simple causal theory of natural kinds, there are a number of ways in which there can be degrees of naturalness among kinds. First, causal relations between properties can be nondeterministic, as noted in considering several of the case studies I have examined. For example, in liquids, viscosity generally decreases with an increase in temperature, roughly because the chemical bonds are "loosened," facilitating liquid flow. But this causal generalization has important exceptions (as noted in section 3.4), notably the element sulfur, whose viscosity increases at a certain temperature between its melting and boiling points. That is because sulfur polymerizes at that temperature, rendering the chemical bonds stronger and the liquid more viscous (though at yet higher temperatures its viscosity decreases again). Generally, causal laws in the special sciences do not hold strictly but are hedged with *ceteris paribus* clauses and admit of exceptions. Those natural kinds associated with stricter causal laws may be deemed more natural because their associated properties are more reliably projectible, while those more riddled with exceptions may be regarded as less natural.

Another aspect of the causal networks associated with natural kinds that admits of degrees has to do with the density of the causal relations involved. Some natural kinds are associated with sets of properties that lead to the manifestation of a host of other properties, which in turn lead to others. But others are fairly limited in the number of properties that they give rise to and cannot be represented as a complex branching network with many nodes but rather as a much simpler network with a limited number of nodes. Natural kinds with a denser branching network are such that there are more things to be discovered of them, and the properties that they give rise to are more varied. Members of the natural kind *lithium*, for example, enter into chemical reactions, are involved in nuclear interactions, and even feature in neuropsychological causal processes. It could be said that the natural kinds with denser causal networks are more natural than those that do not have such a dense network. This dimension of naturalness seems independent of the previous one, since the nature of the causal links themselves may be strict or not, without affecting the density of those links. But these two dimensions of naturalness do not threaten the whole endeavor of identifying natural kinds, since it is not as though all properties have the requisite features to some degree. Some properties, even properties that might be considered natural in their own right do not correspond to natural kinds, nor are they among the primary properties of a natural kind. In other words, some properties are causal "dead ends" and do not give rise to new causal relations either singly or when instantiated with other properties, even if one allows for nonstrict and nondense causal links. For example, having a *melting point of 180.5 °C* does not, according to our current state of knowledge, lead to the instantiation of other causal properties, nor does the co-instantiation of this property with the property of having *viscosity of 1.002 cP* lead to the instantiation of a host of other properties (though both properties are themselves associated with natural kinds, namely, *lithium* and *water*, respectively). Assuming this is an allowable combination of properties, we have no reason to think that their instantiation together would lead regularly to other properties being instantiated.

I have mentioned two other dimensions of naturalness that seem less important because they are even less capable of being measured or assessed. The first has to do with the range of background conditions against which any given natural kind manifests the properties associated with it. A natural kind like *lithium* will demonstrate properties like density or hardness in a wide range of contexts, while a natural kind like *virus* can only demonstrate its capacity to replicate if it locates an appropriate host cell. This difference may lead us to judge that the former is more natural than the latter, though it is not clear how to compare the range of contexts directly. This difference is closely related to a difference of degree that I noted in the discussion of natural law: the degree of stability of the conditions under which laws of nature hold (section 3.5). Another dimension of naturalness has to do with how prevalent members of the kind are in the universe. By this, I do not mean something like the capacity of a natural kind to manifest its properties in different regions of the universe, since this is an idea that I have already criticized in the discussion of natural law (section 3.5). Rather, I mean something far more mundane – namely, the ubiquity or rarity of members of the kind in the universe, either synchronically or across the entire history of the universe. If there are more atoms of hydrogen than atoms of plutonium, or more virions than dogs, in the history of the universe, then that might lead us to view the first member of each pair as being more natural than the second. It is not obvious that such judgments are warranted in general, but they may influence our verdicts concerning the naturalness of kinds.

To recapitulate, I have admitted that there may be degrees of naturalness among natural kinds, but I have also maintained that some kinds are not natural kinds at all, such as categories associated with properties that are causal "dead ends." Even though such properties might themselves be considered natural, as, for example, by "sparse" theorists of properties (Lewis 1983), they do not correspond to natural kinds. Such properties may be among what I have called the secondary properties associated with natural kinds, but they would not be among the primary properties of natural kinds. This is clearly an empirical matter, since there seems to be nothing formally to distinguish real properties that do not correspond to natural kinds from some of the primary properties associated with natural kinds (e.g., having *atomic number 3* or having *constant viscosity with a change in the applied force*, as opposed to having *melting point of 180.5 °C*, or *viscosity of 1.002 cP*). Throughout this book, I have also flagged a number of other examples of categories that do not correspond to natural kinds, such as *hysteria, aquatic animal, aquarium fish, lung cancer, mild cognitive impairment* (MCI), and *ADHD* (construed as a disjunctive category, as contrasted with *ADHD-HI*). Some of these categories even feature in the discourse of science, but they do not do so as projectible categories or as explanatory kind-concepts. Some of them correspond to single properties while others correspond to the co-instantiation of a number of properties. But none of them are associated with causal networks in which a number of other properties are also instantiated.

Once we allow for degrees of naturalness among kinds, this raises the specter of a slippery slope argument. If naturalness admits of degrees along more than one dimension, then perhaps there is no real difference between the natural kinds and the nonnatural kinds, after all. A related view has been expressed by Hacking (2006, 6), who concludes that "Some classifications are more natural than others, but there is no clear and distinct class of natural kinds, and there is no useful vague class either." I agree with Hacking that there is no clear and distinct class but disagree that there is no useful vague class. Just as there are perfectly useful fuzzy kinds across a host of sciences and in a variety of domains, the class of natural kinds itself is fuzzy, yet that is not to say that it is not useful and does not serve a purpose. I have already said what I think that purpose is – namely, distinguishing those categories that identify real divisions in nature from those that do not. Scientists may be quite content to make such judgments without philosophical meddling, but their taxonomic practices tend to be discipline-specific, and I have tried to provide an overview that combines insights from various disciplines. In addition, it is also useful to reflect on the variety of natural kinds in the sciences to disabuse ourselves of some of the essentialist doctrines that have been influential in the recent as well as the more distant history of philosophy. Finally, there are times when the validity of scientific constructs becomes an issue for the general public (e.g., as with *child abuse* or *ADHD-HI*) and when the rationale for positing some categories rather than others is worth making explicit. For all these reasons, it seems to me that the notion of *natural kind* continues to play a useful role.

6.4 REALISM AND PLURALISM

In the previous section, I considered briefly Hacking's skepticism concerning the utility of the class of natural kinds. Some philosophers are far more skeptical than Hacking, holding not that the class of natural kinds is vague or useless, but that it is completely illusory. From their point of view, the categories that we identify are arbitrary and relative to our narrow purposes. For such philosophers, virtually any predicate that groups together a set of individuals in a gerrymandered fashion could be said to be natural from some perspective or relative to some narrow purpose. Antirealism about kinds is expressed eloquently by Rorty (1999, xxvi):

we describe giraffes in the way we do, *as* giraffes because of our needs and interests. We speak a language which includes the word 'giraffe' because it suits our purposes to do so. The same goes for words like 'organ,' 'cell,' 'atom,' and so on – the names of the parts out of which giraffes are made, so to speak. All the descriptions we give of things are descriptions suited to our purposes. No sense can be made, we pragmatists argue, of the claim that some of these descriptions pick out 'natural kinds' – that they cut nature at the joints. The line between a giraffe and the surrounding air is clear enough if you are a human being interested in hunting for meat. If you are a language-using ant or amoeba, or a space voyager observing us from far above, that line is not so clear, and it is not clear that you would need or have a word for 'giraffe' in your language. More generally, it is not clear that any of the millions of ways of describing the piece of space time occupied by what we call a giraffe is any closer to the way things are in themselves than any of the others.

In this passage, Rorty seems to be making two points against a realist view of natural kinds. The first is that categories are picked out for certain purposes, and in our case these purposes are narrowly aligned with our parochial interests. They therefore reflect our own predilections rather than representing features of the universe itself. The second is that our position in the universe or our perspective on reality gives us a narrow view of the world and influences us in drawing lines around our categories. I will try to say something about each of these points in turn.

The idea that our interests influence our categories and that our categories do not, therefore, capture real divisions in nature is one that we have come across in earlier chapters. In section 2.3, I distinguished epistemic purposes from other purposes, arguing that epistemic purposes are privileged since they aim to secure knowledge of real features of the universe, which I elaborated on in terms of real causal patterns.[6] When we are guided by epistemic purposes, our categories will tend to correspond to natural kinds, thereby forging a link between epistemology and metaphysics. I allowed there that this was not always the case in scientific theorizing, and we have seen how nonepistemic purposes can intrude on our attempt to demarcate our categories. Sometimes this is surreptitious, as with the attempt to classify some behaviors by pregnant women as *child abuse* (section 4.7). At other times, it is more overt, as with the explicit statement by some psychiatrists that *mild cognitive impairment* (MCI) ought to continue to be recognized as a psychiatric condition even though it is a heterogeneous category, because of the ethical benefits of attending to the needs of geriatrics with psychiatric problems. But when we alter the borders of our categories in response to nonepistemic interests, these modifications can be detected and the purposes behind them can be ascertained. For example, in the case of the claim that *ADHD* has been considered a single neuropsychological disorder to serve the financial interests of certain pharmaceutical companies, we saw that if the claim is correct, that attempt has not been entirely successful and has been resisted by many scientific researchers, who have instead posited two disorders instead of one. The effort to resist financial, legal, moral, and other purposes from influencing the way in which we categorize natural and social phenomena may not always be successful. But it represents an implicit recognition that these purposes are separate from epistemic purposes and should not be confused with them. As long as the effort to distinguish epistemic from other purposes is not futile, I will continue to maintain that, unless there are nonepistemic purposes that are either overtly or covertly involved in delimiting a category, adopting those methods that serve our epistemic purposes should enable us to identify natural kinds.

---

[6] This claim will not be accepted by many antirealists, but it is beyond the scope of this work to provide a comprehensive response to antirealist positions such as instrumentalism or constructive empiricism.

Rorty's second objection to the claim that some of our categories correspond to natural kinds seems to be that factors such as our size, position in the universe, orientation, and so on play a role in determining which categories we devise. Far from being categories that any observer of the universe would come to, our categories are determined by our idiosyncratic perspective. For example, he says that *giraffe* is not a category that an ant or amoeba would have devised. However, if our scientific categories were entirely determined by our size and position in the universe, then we might not be expected to have come up with such categories as *top quark*, or *cosmic microwave background radiation*, or, indeed, *amoeba*. We have at least managed to transcend some of our limitations and to devise numerous categories that require us to go beyond the limits of our spatial position in the universe, temporal position in the history of the universe, size, and life span (or even that of our species).

A more general version of the objection phrased by Rorty has been called the problem of "selective representation." Mandik and Clark (2002) respond to an objection to this effect, which is articulated by Chemero (1998), who motivates it by asking us to reflect on the world of the lowly tick. In the world represented by this species of organism (sometimes referred to as an *Umwelt*), there is only butyric acid, pressure, and temperature changes. When one considers the tick's limited *Umwelt*, it is hard not to conclude that we may be to the tick as some other cognitive agents may be to us, thus suggesting that our own representations are crude distortions. Since each organism represents the world as containing only those things that matter to its own survival, it would be naive to think that that organism's representations are reflections of the nature of reality.

Mandik and Clark (2002) point to two problems with this argument for antirealism. First, they say that there is a crucial difference between claiming that the organism constructs only a *partial representation* of reality and maintaining that the organism comes up with a *misrepresentation* of reality:

It is one thing to say that ticks represent only X, Y and Z. It is an entirely different thing to say that ticks represent the world as having only X, Y, and Z. The latter case is what is needed for the tick's representations to be in conflict with ours. But the former case is all that the thesis of selective representing is committed to, and the former case is consistent with realism. (Mandik and Clark 2002, 386)

The fact that they are only sensitive to butyric acid, temperature changes, and pressure changes does not show that ticks represent the world as *only* having those features, or as *not* having other features. That is, ticks do not have a representation whose content is that the world does not have other features (even if they could construct such a sophisticated representation, which includes the concept of negation). Another problem with using the phenomenon of selective representation to advance the cause of antirealism has to do with the fact that the function of pursuing our survival in a particular ecological niche is not necessarily incompatible with the function of accurately representing reality. Proponents of selective representation are right to point out that our cognitive systems (and those of all other organisms) were not designed to perform the function of accurately representing reality or of recognizing real causal patterns.[7] Yet, it does not follow that these cognitive systems could not also perform another, compatible function. Finally, Mandik and Clark (2002, 388–389) reply intriguingly to this argument by pointing out that even stating the argument presupposes that there are such things as ticks and that they are responsive to a certain set of objective natural kinds and real properties (*butyric acid, temperature,* and *pressure*). Without such a presupposition, it is unclear how one would even get the selective representation argument for antirealism off the ground.

This answer to antirealism or relativism about kinds in turn may be thought to lead to another objection. Even though humans represent a much vaster set of natural kinds than ticks, what justifies *us* in thinking that the world contains those kinds *exclusively*? There are surely a vast number of kinds that have causal efficacy, undoubtedly more than those uncovered by our current science. I have said that we have managed to transcend many of our limitations as a species. Locke thought that it was unlikely that we would ever be able to know the "corpuscles" of matter (recall section 2.2). Not only have we managed to do so in the intervening three centuries but we have also identified an array of natural kinds in domains that were thought to be beyond our ken just a few decades ago. But will we be able to transcend all such limitations and discover all the kinds there are? There is, of course, no guarantee that we will identify them all; indeed, there may be too many for us ever to identify. If so, does this

---

[7] There is an idea, articulated perhaps most famously by Quine (1969, 126), that we are adept at identifying natural kinds because evolution shaped us to do so, since recognizing "functionally relevant groupings in nature" would have been adaptive. But this line of thought has come under strong criticism, notably from Stich (1990), on the grounds (roughly) that accurate representations are not always adaptive and that evolution is a satisficer not an optimizer. Kitcher (1992, 92) finds both lines of argument overly speculative given our present state of knowledge. I will not try to wade into this debate, since I do not think that one needs anything like Quine's strong claim to advocate realism about natural kinds.

picture lead to too many natural kinds? Pluralism about kinds is one thing, but on this account there is no upper limit on the number of natural kinds that exist in the universe. Though this may appear at first to undermine the notion of natural kinds that many philosophers have, I would argue that it fits better the contemporary scientific conception of the universe. We no longer think of the universe as a single domain of objects within a few orders of magnitude of the size of a human being, and so there is no reason to think that we will be able to fathom all domains and track all causal processes. Unless we are misguidedly anthropocentric, we should not expect that the natural kinds in the universe will be captured by a manageable set of categories that can be easily enumerated by human beings, nor that they will be exhausted by the actual categories that we end up identifying. Hence, an important proviso to the realist position on natural kinds that I have been arguing for is the following: The claim that scientific categories correspond to natural kinds need not imply that *all* natural kinds will be successfully enumerated, even at the end of (human) inquiry. Though the natural kinds correspond to all and only the categories that are epistemically efficacious, we might not successfully discover all such categories in the course of our scientific investigations. The natural kinds that science posits are identified on the basis of their involvement in real causal patterns, but there is no guarantee that the patterns identified by science (even a completed science) are the only real causal patterns in the universe. There may be no way of knowing how many others lie out there uncharted.[8] Realists about natural kinds may need to content themselves with the *truth* and *nothing but the truth* without also insisting on the *whole truth*.[9]

The realization that there may be many more natural kinds than some philosophers might have expected goes hand in hand with a recognition that contemporary science reveals a world of multiple domains, each characterized by distinct causal processes. When I articulated the notion of a *domain* in section 3.6, I contrasted it with the conception of *levels*, which has been quite prevalent in recent philosophy (though it has also had its share of critics). While acknowledging the difficulty of demarcating domains, I have proposed that they can be captured in terms of two

---

[8] Compare Ladyman and Ross (2007, 203): "Thus there are (presumably) real patterns in lifeless parts of the universe that no actual observer will ever reach, and further real patterns whose data points are before our eyes right now, but which no computer we can instantiate or design will ever marshal the energy to compact."

[9] But this position is less pluralist than some self-described realist views, notably "promiscuous realism" (Dupré 1993), as seen in section 2.3.

dimensions: spatiotemporal and aspectual. This characterization of domains is admittedly vague, but I think it is preferable to talk of levels for two reasons. First, there is no strict hierarchy of domains; even though there may be an ordering on the spatiotemporal dimension, there does not seem to be a way to rank aspects. For instance, when it comes to tracking the causal processes in which nuclides are involved, the chemical and nuclear processes are not ranked, and the properties of atomic number and mass number are not ordered in any way. Second, because of the phenomenon of crosscutting kinds, domains are not always compositionally related to one another in a straightforward way, with the entities of domain D being composed of the entities of D', and so on. That is because the individuals that possess the macrolevel properties are not generally simple mereological sums of the individuals that possess the lower-level properties. In many cases, the macro-level entities are individuated extrinsically, functionally, or even etiologically, and cannot be identified with simple aggregates of microlevel entities. Even domains at the same spatiotemporal scale, which involve different aspects of the same phenomena, are not always populated by the same entities. For instance, it may seem as though chemical and nuclear causal processes apply to the same individuals, but nuclear processes pertain to the nucleus of the atom while chemical processes concern the atom as a whole including its electron orbitals. Since domains are also aspectual, some of them individuate entities extrinsically or functionally, in such a way that cannot be captured by a (single) theory from a different domain. The biochemical, for example, is not just the chemical aggregated. Biochemical entities may be individuated functionally in terms of their effects on the organism or the role that they play in a particular system.

6.5 MIND-INDEPENDENCE AND SOCIAL CONSTRUCTIONISM

In an essay entitled, "The Naturalists Return," Kitcher (1992, 104) characterizes realism (in a naturalist context) as follows:

Minimal realism holds that there are objects independent of human cognition. Strong realism adds the thesis that, independently of us, these objects are assorted into natural kinds and that there are causal processes in which they participate. The task of science is to expose the causal structure of the world, by delineating the pre-existent natural kinds and uncovering the mechanisms that underlie causal dependencies.

It might seem as though, on Kitcher's definition, the naturalist account of natural kinds that I have been developing is "strongly" realist. But as he defines it, strong realism includes the thesis that natural kinds are "independent of us," presumably meaning independent of human beings and "human cognition." However, this is an aspect of traditional realism that I have explicitly repudiated, for reasons that I spelled out in section 4.5. To recapitulate, this criterion would rule out all psychological and social kinds as nonreal. More importantly, defenders of the criterion (e.g., Boyd 1989) have not articulated a clear way of distinguishing a problematic form of dependence (e.g., constitutive) from a nonproblematic form (e.g., causal). Perhaps it appears as though I have given up on this attempt too quickly and there may still be a way of showing that social kinds can be dependent on human minds and mental activity in one way but independent of them in another (so that they do not have to be disqualified from being natural kinds). But I also suggested that no matter how one tried to spell out the distinction between causal and constitutive (or conceptual) mind-dependence, mental kinds would surely come out to be constitutively mind-dependent. Take kinds like *Alzheimer's disease* or *ADHD-HI*, which are at least plausible candidates for being natural kinds. These are psychological or psychiatric kinds whose very existence is bound up with mentation. They are surely constitutively or conceptually mind-dependent, if anything is, and so would seem to be doomed to end up on the wrong side of the divide. One might be able to save some psychological or mental kinds if one were a type reductionist about the psychological, by deeming them to be neural rather than mental kinds, and hence not constitutively mind-dependent. But that seems like a very high price to pay for the sake of making a distinction that is not needed in the first place, as I will now try to show.

Having suggested that any attempt to make the distinction between causal and constitutive mind-dependence will encounter a difficulty with psychological kinds, I will argue that the whole effort to make such a distinction is misguided in this context because mind-dependence is a red herring. There are many things in the world whose existence is dependent on human beings, their minds, and their actions, including many kinds of things: *marriage, government, recession, alcoholism*, and *child abuse*, but also *polyethylene, Canis familiaris*, and *hypertension*. Some of these are social kinds but others are not. Humans and their minds are part of the natural world and they feature in causal processes, sometimes giving rise to or enabling certain kinds of things to come into existence, whether unintentionally and without awareness (e.g., *alcoholism, recession*), or deliberately and by design (e.g., *polyethylene, government*). I have argued that to rule out the categories that humans give rise to, even those that they deliberately engineer, would be to dismiss real causal patterns. Instead, in identifying bogus categories that do not correspond to natural kinds, what we want to rule out are not kinds that are mind-dependent but ones that have no basis in reality. After all, if there were a mechanism by which we could influence the world directly through our beliefs, those mental influences would be real and the changes would be ones that we would have to take into account in our scientific inquiries. The problem with bogus kinds is not that they are mind-dependent, but that they are world-independent, as it were. To say that what they have in common is mind-dependence is to offer a proxy for what is really objectionable.

So what is really objectionable? If we want to be able to accurately reflect causal patterns in an objective manner, then I would submit that what we want to avoid is subjectivity in the sense of biases that subvert our epistemic purposes. In previous chapters, we have seen some instances of categories that were modified, were attempted to be modified, or were alleged to be modified by humans, whether deliberately or not, in ways that tended not to reflect real causal patterns. In all such cases, I argued, epistemic purposes were sacrificed (or would have been sacrificed) in favor of other purposes. Nonepistemic interests have the potential to sidetrack investigators from uncovering real causal patterns and push them in the direction of redrawing the boundaries of categories in such a way that they cease to track causal processes or networks.

The mind-dependence of categories is not the problem, but rather the pursuit of nonepistemic purposes instead of epistemic ones. Therefore, we should not ask that our categories be mind-independent, but just that they serve epistemic purposes and aim to reflect real causal patterns. Objectivity when it comes to natural kinds is a matter of being guided by epistemic purposes; mind-independence is just a diversion. Of course, there is a great deal more to be said about what epistemic purposes are and what it is to be guided by them, and philosophers of science have made substantial contributions to these questions throughout the modern era from Bacon's "Four Idols" onwards. But an account of scientific methodology is well beyond the scope of this particular inquiry, and I have taken it for granted in this account of natural kinds. If that is granted, then one way to formulate realism about natural kinds is as follows: Natural kinds divide the world into individuals that share causal properties, enter into the same or similar causal relationships, and give rise to the same or similar causal processes. Scientific categories that correspond to natural kinds are therefore projectible, enabling us to infer the presence of some properties from other properties on the basis of the causal relations between them. Natural kinds need not be mind-independent, since some of those causal processes may involve humans and their mental states, but in order to identify them we must be guided by epistemic purposes and not be deflected by non-epistemic interests.

There is one further concern regarding the alleged arbitrariness of category boundaries that may raise fears of antirealism about kinds. I have argued on several occasions that there can be kinds with fuzzy boundaries, with members that are indeterminate between two kinds or members that are marginal (as opposed to focal) instances of a kind. Scientists can treat these cases in one of two ways. They can either draw a precise line between the two kinds or they can let their categories reflect the fuzziness that appears in nature. Consider a kind K associated with a property that varies continuously along some dimension D and suppose that focal instances of K have $D = n$. Then we can either rule that all members must fall within a range of $D = n \pm \delta$ and that anything outside that range is not a member, or we can hold that there is graded membership in the kind, with focal members falling within that range, marginal members in the wider range of have $D = n \pm 2\delta$, and so on. Thus, we can either posit a precise cutoff between members and nonmembers, or we can posit focal and marginal members. Both courses of action are attested to in scientific practice, but scientists may also live with the ambiguity and may not always make a clear choice. However, in the first case, scientists may be said to be making a clear-cut distinction where there is no sharp difference in nature. This may be taken as a sign that our scientific categories sometimes distort the nature of reality and impose divisions on nature rather than merely reflect the divisions that exist in nature. There is something to this complaint, but it should not be overstated. I have already mentioned that the category *polymer* is not neatly demarcated but has fuzzy boundaries (chemists sometimes distinguish polymers from oligomers). Similarly, in the medical sciences, one oft-cited example in which a precise cutoff has been posited is the category *hypertension*, which is defined as a condition in which a person's blood pressure is persistently ≥140/90 mmHg. This may appear arbitrary and the reflection of human stipulation rather than a boundary that exists in nature. But the reason for drawing the line there rather than elsewhere does not depend simply on the whim of medical researchers. It reflects the fact that investigations of human health and morbidity have concluded that the risks of developing other medical conditions rise substantially beyond that point. Causal relations in nature are often not deterministic and the sciences need to reflect that fact. Though there is a type of distortion involved in demarcating the boundaries of fuzzy kinds, it is not a clear vindication of antirealism. This is not a matter of inventing kinds out of whole cloth. Instead, it is a matter of representing the boundaries of a natural kind as being more precise than they actually are. Scientific inquirers often have some leeway in deciding how to draw the boundaries around fuzzy kinds, but I do not think that this should occasion wholesale antirealism about fuzzy kinds, much less natural kinds in general.

I think it is worth relating these ideas about mind-dependence to the very widespread (at least outside of philosophical circles) conviction that the categories we devise in science and, indeed, the kinds we purport to discover are "socially constructed." A claim to the effect that a category or kind, like *quark* (Pickering 1984), *ocean* (Steinberg 2001), *climate change* (Pettenger 2007), *sexuality* (Seidman 2010), or *urban schooling* (Miron 1996), is socially constructed is not a straightforward one.[10] In order to make sense of such claims, it is worth considering two versions, as applied to the *category* (or kind-concept) and as applied to the *kind* itself (and its members). In each case, it is also worth considering the import of the claim as applied to *social* kinds and categories, and as applied to *nonsocial* kinds and categories.

First, consider the claim of social construction as it pertains to categories. All our categories and concepts are of our own making; though some may be rooted in innate cognitive capacities, all are in some sense a product of human mental effort. This mental effort is hardly ever wholly the result of individual exertion but involves social cooperation to some extent. Hence, the claim that categories or kind-concepts are socially constructed, in the most literal sense – namely, that they are devised by human beings collectively in concert with others – is a fairly trivial claim. This is true of social categories no less than of nonsocial categories. But one might say that the point is that the social influences on these categories are more profound than is widely suspected and that these categories are strongly shaped by social factors. It seems to me that this claim can be understood in terms of the ideas I have been discussing concerning the ways in which nonepistemic interests can influence categories in the sciences, especially the social sciences. If that is the upshot of the claim that a certain category is socially constructed – namely, that it has been significantly shaped by nonepistemic social, political, or other interests –

---

[10] Here I have taken a leaf from Hacking (1999), who lists some two dozen items that are said to be socially constructed according to recent book titles; my list is also drawn from recent book titles, though only one of them (*quark*) also appears on Hacking's list.

then that is a substantive claim and one that should be examined on its merits, whether in the social or nonsocial sciences. I think that this is often what is meant by saying that, say, *climate change* or *ADHD* is socially constructed: These categories have allegedly been shaped by nonepistemic interests and reflect certain biases deriving from political, financial, legal, or moral purposes that humans might bring to their inquiries into the natural and social realms. I cannot pretend to do justice to the range of contributions that have been made to the social constructionist enterprise, but many such contributions have painstakingly traced the way in which political and other influences have contributed to shaping categories in the natural and social sciences. These studies often engage in a research strategy that involves "unmasking," or tracing covert influences behind the formulation of a category. At least that is one way to interpret the project of social constructionism. No doubt many social constructionists would disavow the kind of distinction that I have been making between epistemic and other interests, insisting that this is a false dichotomy and that there is no such distinction to be made.[11] Hence, they would reject this interpretation of the social constructionist claim about categories. In any inquiry, they might say, there are just competing "wills to power," and there can be no such thing as an objective scientific investigation that serves epistemic interests alone. When social constructionists show that a certain inquiry is motivated by political interests, they may also add that political interests of one kind or another are inevitable. I will not try to take on this issue here or further justify my interpretation of social constructionism about categories. But despite my disagreement with this point, I nevertheless think that many such studies effectively unmask social influences on category construction that might otherwise remain unnoticed. Hence, there is often merit to the assertion that a particular scientific category is socially constructed – namely, when it is at least partly shaped by nonepistemic interests.

As for the claim that a kind itself may be socially constructed, that claim admits of a problematic and a nonproblematic interpretation. When applied to a kind like *urban schooling*, which is a social kind in the sense that it applies to human beings and involves interactions between them in a social setting, the claim may seem to be trivially true. To say that *urban*

---

[11] I am thinking of the influential slogan, "knowledge is power," frequently attributed to Michel Foucault. Gutting (2011) explicates this Foucauldian idea as follows: "Foucault's point is . . . that, at least for the study of human beings, the goals of power and the goals of knowledge cannot be separated: in knowing we control and in controlling we know."

schooling is socially constructed is akin to saying that a cell is biologically constructed. The practice of schooling in an urban setting is clearly influenced by members of society, social institutions, and a range of social forces. If so, a claim that it is socially constructed is hardly worth making because it is obviously true. However, what social constructionist studies often undertake is a detailed examination of the processes by which these social kinds have evolved and become entrenched. Hence, the interest is not in the bare claim itself but in the way in which sociologists, social historians, and others elaborate on the claim and their meticulous accounts of the social forces that are involved in shaping the social kind in question. By contrast, a claim to the effect that a kind such as *quark* is socially constructed is more questionable. Since social forces do not seem to have any causal influence on the processes in which quarks are implicated nor are they influenced directly by them, the claim may seem obviously false. There may be purported kinds in the natural sciences for which it is true that they are in fact shaped by social forces, but it is at the very least a far-fetched claim. Therefore, for both clear-cut social kinds and unambiguous nonsocial kinds, the claim of social construction is problematic because it seems to be obviously true and highly improbable, respectively. However, when it comes to a kind like *gender*, the claim may be more interesting. The reason is that *gender* and some other kinds (notably *race*) occupy an interesting position that straddles the biological and social sciences. There are a number of kinds that are not unambiguously social about which the claim of social construction is a substantive and interesting one. In these cases, social constructionists often alert us that though these kinds may commonly be thought to be biological and to pertain to one or more biological domains (e.g., genetics), in fact they pertain to one or more social domains (e.g., sociology) (cf. Haslanger 2006). Exactly what it is for a kind to be social as opposed to biological, and how one ascertains this, are important issues that I will not attempt to pronounce on. In the case of *gender*, this task is closely related to the distinction that is often made between *gender* and *sex*. In the case of *race*, it would involve showing that there is no such natural kind to be identified in the domain of evolutionary biology or genetics, but rather that such a kind exists as a result of prevailing attitudes and prejudices in certain societies.[12]

---

[12] The issues here are involved and have been hotly debated. For a nuanced account claiming that races are both socially and biologically constituted, see Kitcher (1999); for a notable argument that *race* is a biological kind, see Andreasen (1998).

There is an important difference between the social constructionist claim regarding a category and that regarding a kind. When it comes to a category, the claim can be interpreted as a statement to the effect that it has been (illegitimately) influenced by nonepistemic interests, and each such claim needs to be examined on its own merits. When it comes to a kind, such assertions are trivial if the kind is overtly social, and questionable if the kind is not. However, the most interesting claims of this type involve kinds that straddle the social and nonsocial realms. In these cases, social constructionists have often shown that a kind that has been widely thought to be biological is actually sociological in terms of its causes and effects.

In this section, I have tackled two claims of human-dependence concerning categories and kinds. A prevalent view about kinds is that they must be mind-independent to be real, but I argued that this claim is beside the point. Not only has the relevant (noncausal) independence claim not been articulated adequately by philosophers, but also mind-independence is irrelevant to realism since what we want from our kinds is not that they should be independent of the human mind but that they should be instantiated in the universe (which includes human minds) and participate in real causal patterns. A more specific version of the mind-dependence thesis pertains to the claim of social construction. The social constructionist claim regarding categories is trivially true, though it is sometimes instructive to have an account of the process whereby our categories have been constructed. As for the social constructionist claim about kinds themselves, it is vacuous with regards to social kinds and implausible concerning most nonsocial kinds. But there is a plausible version of the thesis with regards to those kinds that are commonly thought to be biological but are instead (or in addition) social kinds (e.g., *race*).

6.6 CONCLUSION

This naturalist inquiry into natural kinds suggests that there is considerable uniformity across scientific domains when it comes to the characterization of natural kinds. Insofar as the kinds that they discover are concerned, I have found little reason to think that there is a divide between the sciences that study microscopic and macroscopic phenomena, or between the physical and chemical sciences and the biological sciences, or between the natural sciences and the social sciences. Many philosophers, even anti-essentialists about natural kinds, tend to believe that the kinds of physics and chemistry are "essence kinds" while those of biology and other sciences are "cluster kinds" (Chakravartty 2007), but that is not borne out by a detailed examination of some of the kinds on both sides of this divide. It is not the case that some types of natural kinds figure solely in macroscopic and not in microscopic domains (e.g., polythetic kinds, fuzzy kinds, crosscutting kinds), or only in the biological sciences and not the physical and chemical sciences (e.g., etiological kinds, copied kinds), or just in the social but not the nonsocial sciences (e.g., interactive kinds). Furthermore, in all cases, I have argued that there is little heterogeneity among natural kinds across disciplines and that in all cases natural kinds can be characterized as nodes in causal networks.

In saying that naturalism points us in the direction of a unitary account of natural kinds that applies across scientific disciplines, I am not denying that there is an obvious asymmetry between the basic and special sciences, or between the domains of the smallest constituents of the universe and other domains. There are at least three kinds of asymmetry, which are not entirely independent. Most obviously, there is an asymmetry of constitution, in that the entities investigated in all other domains are literally composed of those in the most fundamental physical domain. Chemical compounds, biological organisms, and human societies are made up of nothing but quarks, leptons, and bosons. This does not mean that the properties and natural kinds of these domains are nothing but the properties and natural kinds of the domain of elementary particle physics. The constitutive relation applies to entities but it does not literally apply to properties and kinds. Moreover, even the constitutive claim concerning entities obtains only loosely given the crosscutting thesis, since macroentities are individuated differently from microentities (see section 6.4). Similarly, as Kornblith (1993, 54–55) observes, if we remove a single atom from a dog, it would no longer be the same collection of atoms but it would (probably) be the same dog. Even though a dog is nothing but a collection of atoms, it does not follow that *this* dog is identical to *this very* collection of atoms.

The second kind of asymmetry is that of supervenience. I have avoided a detailed examination of the concept of supervenience in this book, but I did argue that the only supervenience compatible with modern scientific taxonomy is global rather than local. This means that there is no difference in all other domains without a difference in the domain of elementary particles, even though the difference may not be localized in the same vicinity. If a virus has infected a host cell or if a person has changed her mind, then we can be certain that some quarks and leptons must have altered their positions in space-time or their momenta or both. But the opposite is not true: There can be changes in the states of elementary particles without changes that show up in the domain of virology or psychology, or in any other domain for that matter. Relatedly, the laws and generalizations applicable to the elementary particles place constraints on all other domains, but the converse is not generally true. For instance, the law of the conservation of mass-energy constrains natural selection, but the economic laws of supply and demand do not constrain elementary particle interactions. At the same time, we should not exaggerate these kinds of constraints, because they are often rather minimal. Global supervenience is a rather weak version of the supervenience claim. Many of the laws that describe the behavior of elementary particles simply do not apply to, or are evened out in, other domains. Quantum effects seldom show up in biology and never in social processes, which proceed more or less as though there were no such effects.

The third type of difference between the domain of elementary particles (as well as some of the other microscopic domains) and other domains is a little harder to state with precision. But it is reflected in the kinds of natural laws and empirical generalizations that are found in these respective domains. As we saw in section 3.4, the generalizations of the special sciences admit of more exceptions than at least some of the generalizations of elementary particle and nuclear physics. In the smallest domains, there would seem to be fewer causal networks to track. It is strictly false that all atoms of lithium are the same (because of isotopes and ions), but there is less variety in them than there is in cancer cells. Hence, there are fewer exceptions to the generalizations we can make about them. Though size and complexity are not directly proportional, it appears as though elementary particles have a kind of simplicity that few other kinds do. As we move towards the smallest spatiotemporal scale, there are fewer exceptions and they are better known. There also seem to be fewer degrees of freedom at the shorter end of the spatiotemporal spectrum, in the sense that there are fewer allowable combinations of causal properties and hence fewer natural kinds. Therefore, in some rough sense, there may be fewer natural kinds and we may come to know a greater proportion of them. There are surely fewer possible chemical elements than there are possible chemical compounds, and fewer possible chemical compounds than there are possible biological species. But that does not make the natural kinds at smaller spatiotemporal scales fundamentally different from those at larger ones, and it should not lead us to conclude that the only natural kinds in the universe are those in the domain of elementary particles.

Some philosophers of a microphysical fundamentalist persuasion may remain unconvinced by my arguments and may continue to insist that the universe consists of "nothing but" quarks, leptons, and bosons, and hence that the only natural kinds, properly speaking, are the kinds of elementary particles (e.g., *top quark, neutrino*). (To draw the line anywhere else seems inconsistent since protons are surely nothing but quarks, and so on.) But even so, this would leave open the question of what distinguishes the categories identified in all the other scientific domains from any number of artificial, arbitrary, or gerrymandered collections of individuals that one might care to conjure up. That is a significant philosophical question in its own right and one that cannot be lightly dismissed. If some philosophers still insist on reserving the term "natural kinds" exclusively for the kinds of elementary particles, then the kinds that I have been discussing may instead be called "naturalized kinds."

Bibliography

Andreasen, R. (1998), "A New Perspective on the Race Debate," *British Journal for the Philosophy of Science* 49, 199–225.

APA (American Psychiatric Association) Committee on Nomenclature and Statistics (1968), *Diagnostic and Statistical Manual of Mental Disorders* (2nd edn.) (Washington, DC: American Psychiatric Association).

(1980), *Diagnostic and Statistical Manual of Mental Disorders-III* (3rd edn.) (Washington, DC: American Psychiatric Association).

(1987), *Diagnostic and Statistical Manual of Mental Disorders-III-R* (rev. 3rd edn.) (Washington, DC: American Psychiatric Association).

(1994), *Diagnostic and Statistical Manual of Mental Disorders-IV* (4th edn.) (Washington, DC: American Psychiatric Association).

(2000), *Diagnostic and Statistical Manual of Mental Disorders-IV-TR* (rev. 4th edn.) (Washington, DC: American Psychiatric Association).

Armstrong, D. (1978a), *Universals and Scientific Realism: Nominalism and Realism*, vol. I (Cambridge University Press).

(1978b), *Universals and Scientific Realism: A Theory of Universals*, vol. II (Cambridge University Press).

(1980/1997), "Against Ostrich Nominalism: A Reply to Michael Devitt," in D. H. Mellor and A. Oliver (eds.), *Properties* (Oxford University Press), 101–111.

(1986), "In Defence of Structural Universals," *Australasian Journal of Philosophy* 64, 85–88.

(1989), *Universals: An Opinionated Introduction* (Boulder, CO: Westview Press).

(1992/1997), "Properties," in D. H. Mellor and A. Oliver (eds.), *Properties* (Oxford University Press), 160–172.

(1997), *A World of States of Affairs* (Cambridge University Press).

Ball, P. (2004), *The Elements: A Very Short Introduction* (Oxford University Press).

Bambrough, R. (1961), "Universals and Family Resemblances," *Proceedings of the Aristotelian Society* 61, 207–222.

Barkley, R. A. (1997), *ADHD and the Nature of Self-Control* (New York: Guilford Press).

Batterman, R. (2000), "Multiple Realizability and Universality," *British Journal for the Philosophy of Science* 51(1), 115–145.

Benditt, T. M. (2010), "Normality, Disease, and Enhancement," in H. Kincaid and J. McKitrick (eds.), *Establishing Medical Reality: Essays in the Metaphysics and Epistemology of Biomedical Science* (Dordrecht: Springer), 13–21.

Bird, A. (2008), "Causal Exclusion and Evolved Emergent Properties," in R. Groff (ed.), *Revitalizing Causality: Realism About Causality in Philosophy and Social Science* (New York: Routledge), 163–178.

Block, N. (2003), "Do Causal Powers Drain Away?," *Philosophy and Phenomenological Research* 67, 133–150.

Bogen, J. (1988), "Comments on 'The Sociology of Knowledge About Child Abuse,'" *Noûs* 22, 65–66.

Boyd, R. (1989), "What Realism Implies and What It Does Not," *Dialectica* 43, 5–29.

(1991), "Realism, Anti-Foundationalism, and the Enthusiasm for Natural Kinds," *Philosophical Studies* 61, 127–148.

(1999a), "Homeostasis, Species, and Higher Taxa," in R. A. Wilson (ed.), *Species: New Interdisciplinary Essays* (Cambridge, MA: MIT Press), 141–186.

(1999b), "Kinds, Complexity and Multiple Realization," *Philosophical Studies* 95, 67–98.

Brigandt, I. (2003), "Species Pluralism Does Not Imply Species Eliminativism," *Philosophy of Science* 70, 1305–1316.

Cargile, J. (2003), "On 'Alexander's' Dictum," *Topoi* 22, 143–149.

Carlson, S. M. (2005), "Developmentally Sensitive Measures of Executive Function in Preschool Children," *Developmental Neuropsychology*, 28(2), 595–616.

Cartwright, N. (1999), *The Dappled World: A Study of the Boundaries of Science* (Oxford University Press).

Chakravartty, A. (2007), *A Metaphysics for Scientific Realism: Knowing the Unobservable* (Cambridge University Press).

Chemero, A. (1998), "A Stroll Through the Worlds of Animals and Humans," *Psyche* 4, 1–10.

Cook, M. (1980), "If 'Cat' Is a Rigid Designator, What Does It Designate?," *Philosophical Studies* 37, 61–64.

Cooper, R. (2004), "Why Hacking Is Wrong About Human Kinds," *British Journal for the Philosophy of Science* 55, 73–85.

Craik, F. I. M., Bialystok, E., and Freedman, M. (2010), "Delaying the Onset of Alzheimer Disease: Bilingualism as a Form of Cognitive Reserve," *Neurology* 75, 1726–1729.

Craver, C. (2009), "Mechanisms and Natural Kinds," *Philosophical Psychology* 22, 575–594.

Crawford, D. H. (2000), *The Invisible Enemy: A Natural History of Viruses* (Oxford University Press).

(2011), *Viruses: A Very Short Introduction* (Oxford University Press).

Curtis, C., Shah, S. P., Chin, S. F., *et al.* (2012), "The Genomic and Transcriptomic Architecture of 2,000 Breast Tumours Reveals Novel Subgroups," *Nature* 486, 346–352.

Cronbach, L. J. and Meehl, P. E. (1955), "Construct Validity in Psychological Tests," *Psychological Bulletin* 52, 281–302.

Dennett, D. C. (1991), "Real Patterns," *Journal of Philosophy* 88(1), 27–51.

Devitt, M. (2005), "Scientific Realism," in F. Jackson and M. Smith (eds.), *The Oxford Handbook of Contemporary Philosophy* (Oxford University Press), 767–791.

Donnellan, K. (1983), "Kripke and Putnam on Natural Kind Terms," in C. Ginet and S. Shoemaker (eds.), *Knowledge and Mind: Philosophical Essays* (Oxford University Press), 84–104.

Douglas, Mary (1986), *How Institutions Think* (Syracuse, NY: Syracuse University Press).

Dupré, J. (1993), *The Disorder of Things: Metaphysical Foundations of the Disunity of Science* (Cambridge, MA: Harvard University Press).

  (1999), "Are Whales Fish?," in D. L. Medin and S. Atran (eds.), *Folkbiology* (Cambridge, MA: MIT Press), 461–476.

  (2004), "Human Kinds and Biological Kinds: Some Similarities and Differences," *Philosophy of Science* 71(5), 892–900.

Elder, C. (2004), *Real Natures and Familiar Objects* (Cambridge, MA: MIT Press).

Ellis, B. (2001), *Scientific Essentialism* (Cambridge University Press).

Ereshefsky, M. (ed.) (1992), *The Units of Evolution: Essays on the Nature of Species* (Cambridge, MA: MIT Press).

  (ed.) (2004), "Bridging the Gap Between Human Kinds and Biological Kinds," *Philosophy of Science* 71, 912–921.

  (ed.) (2010), "What's Wrong with the New Biological Essentialism?," *Philosophy of Science* 77, 674–685.

Ereshefsky, M. and Matthen, M. (2005), "Taxonomy, Polymorphism, and History: An Introduction to Population Structure Theory," *Philosophy of Science* 72, 1–21.

Faraone, S. V. (2005), "The Scientific Foundation for Understanding Attention-Deficit/Hyperactivity Disorder As a Valid Psychiatric Disorder," *European Child & Adolescent Psychiatry* 14, 1–10.

Fay, B. (1983/1994), "General Laws and Explaining Human Behavior," in M. Martin and L. C. McIntyre (eds.), *Readings in the Philosophy of Social Science* (Cambridge, MA: MIT Press), 91–110.

Fodor, J. (1974), "Special Sciences (or: The Disunity of Science as a Working Hypothesis)," *Synthese* 28, 97–115.

  (1997), "Special Sciences: Still Autonomous After All These Years," *Noûs* 31, 149–163.

Franklin, F. and Franklin, C. L. (1888), "Mill's Natural Kinds," *Mind* 13, 83–85.

Furman, L. M. (2008), "Attention-Deficit Hyperactivity Disorder (ADHD): Does New Research Support Old Concepts?," *Journal of Child Neurology* 23, 775–784.

Garson, J. (2011), "Selected Effects and Causal Role Functions in the Brain: The Case for an Etiological Approach to Neuroscience," *Biology and Philosophy* 26, 547–565.

Goldstein, S. and Naglieri, J. A. (2008), "The School Neuropsychology of ADHD: Theory, Assessment, and Intervention," *Psychology in the Schools* 45, 859–874.

Goodman, N. (1954/1979), *Fact, Fiction, and Forecast* (Cambridge, MA: Harvard University Press).

Graham, J. E. and Ritchie, K. (2006), "Mild Cognitive Impairment: Ethical Considerations for Nosological Flexibility in Human Kinds," *Philosophy, Psychiatry, & Psychology* 13, 31–43.

Gratzer, W. (2009), *Giant Molecules: From Nylon to Nanotubes* (Oxford University Press).

Greene, B. (2011), *The Hidden Reality: Parallel Universes and the Deep Laws of the Cosmos* (New York: Knopf).

Griffiths, P. E. (1997), *What Emotions Really Are: The Problem of Psychological Categories* (University of Chicago Press).

(1999), "Squaring the Circle: Natural Kinds with Historical Essences," in R. A. Wilson (ed.), *Species: New Interdisciplinary Essays* (Cambridge, MA: MIT Press), 209–228.

(2002), "What Is Innateness?," *Monist* 85, 70–85.

(2004), "Emotions as Natural and Normative Kinds," *Philosophy of Science* 71, 901–911.

Grosberg, A. Y. and Khokhlov, A. R. (1997), *Giant Molecules: Here, There, and Everywhere* (San Diego, CA: Academic Press).

Guala, F. (2010), "Infallibilism and Human Kinds," *Philosophy of the Social Sciences* 40(2), 244–264.

Gutting, G. (2011), "Michel Foucault", in Edward N. Zalta (ed.), *Stanford Encyclopedia of Philosophy* (Fall 2011 Edition), http://plato.stanford.edu/archives/fall2011/entries/foucault/.

Hacking, I. (1986/1999), "Making up People," in Mario Biagioli (ed.), *Science Studies Reader* (New York: Routledge), 161–171.

(1988), "The Sociology of Knowledge About Child Abuse," *Noûs* 22, 53–63.

(1991a), "A Tradition of Natural Kinds," *Philosophical Studies* 61, 109–126.

(1991b), "The Making and Molding of Child Abuse," *Critical Inquiry* 17, 253–288.

(1992), "World-Making by Kind-Making: Child Abuse for Example," in Mary Douglas and David Hull (eds.), *How Classification Works* (Edinburgh University Press), 180–238.

(1995a), *Rewriting the Soul: Multiple Personality and the Sciences of Memory* (Princeton University Press).

(1995b), "The Looping Effects of Human Kinds," in Dan Sperber, David Premack, and Ann J. Premack (eds.), *Causal Cognition: A Multidisciplinary Approach* (Oxford University Press), 351–383.

(1999), *The Social Construction of What?* (Cambridge, MA: Harvard University Press).

(2002), *Historical Ontology* (Cambridge, MA: Harvard University Press).

(2006), "Kinds of People: Moving Targets," British Academy Lecture, April 11 (available at: www.britac.ac.uk/pubs/src/_pdf/hacking.pdf).

Hamilton, W. D. (1964), "The Genetical Evolution of Social Behaviour I," *Journal of Theoretical Biology* 7, 1–16.

Hanahan, D. and Weinberg, R. A. (2000), "The Hallmarks of Cancer," *Cell* 100, 57–70.

(2011), "Hallmarks of Cancer: The Next Generation," *Cell* 144, 646–674.

Haslam, N. (2002), "Kinds of Kinds: A Conceptual Taxonomy of Psychiatric Categories," *Philosophy, Psychiatry & Psychology* 9, 203–217.

Haslanger, S. (2006), "What Good Are Our Intuitions? Philosophical Analysis and Social Kinds," *Aristotelian Society Supplementary Volume* 80, 89–118.

Haslanger, S. A. (1995), "Ontology and Social Construction," *Philosophical Topics* 23, 95–125.

Hawking, S. and Mlodinow, L. (2010), *The Grand Design* (New York: Random House).

Heil, J. (2003), *From an Ontological Point of View* (Oxford University Press).

Hendry, R. F. (2006), "Elements, Compounds, and Other Chemical Kinds," *Philosophy of Science* 73, 864–875.

Hey, Jody (2001), "The Mind of the Species Problem," *Trends in Ecology & Evolution* 16, 326–329.

Horgan, T. (2001), "Causal Compatibilism and the Exclusion Problem," *Theoria* 16, 95–116.

Karbasizadeh, A. E. (2008), "Revising the Concept of Lawhood: Special Sciences and Natural Kinds," *Synthese* 162, 15–30.

Kendler, K. S., Zachar, P., and Craver, C. (2011), "What Kinds of Things Are Psychiatric Disorders?," *Psychological Medicine* 41, 1143–1150.

Khalidi, M. A. (1993), "Carving Nature at the Joints," *Philosophy of Science* 60, 100–113.

(1998a), "Natural Kinds and Crosscutting Categories," *Journal of Philosophy* 95, 33–50.

(1998b), "Incommensurability in Cognitive Guise," *Philosophical Psychology* 11, 29–43.

(2005), "Against Functional Reductionism in Cognitive Science," *International Studies in the Philosophy of Science* 19, 319–333.

(2008), "Temporal and Counterfactual Possibility," *Sorites* 20, 37–42.

(2010), "Interactive Kinds," *British Journal for the Philosophy of Science* 61, 335–360.

Kim, J. (1982), "Psychophysical Supervenience," *Philosophical Studies* 41, 51–70.

(1992), "Multiple Realization and the Metaphysics of Reduction," *Philosophy and Phenomenological Research* 52, 1–26.

(1998), *Mind in a Physical World* (Cambridge, MA: MIT Press).

(2003), "Blocking Causal Drainage and Other Maintenance Chores with Mental Causation," *Philosophy and Phenomenological Research* 67, 151–176.

Kincaid, H. (1990), "Defending Laws in the Social Sciences," *Philosophy of the Social Sciences* 20(1), 56–83.

Kistler, M. (1999), "Multiple Realization, Reduction and Mental Properties," *International Studies in the Philosophy of Science* 13, 135–149.

(2002), "The Causal Criterion of Reality and the Necessity of Laws of Nature," *Metaphysica* 3, 57–86.

Kitcher, P. (1992), "The Naturalists Return," *Philosophical Review* 101, 53–114.

(1995), *The Advancement of Science* (Oxford University Press).

(1999), "Race, Ethnicity, Biology, Culture," in L. Harris (ed.), *Racism* (Amherst, NY: Humanity Books), 87–117.

Kornblith, H. (1993), *Inductive Inference and Its Natural Ground* (Cambridge, MA: MIT Press).

Koslicki, K. (2008), "Natural Kinds and Natural Kind Terms," *Philosophy Compass* 3/4, 789–802.

Kripke, S. (1980), *Naming and Necessity* (Cambridge, MA: Harvard University Press).

Ladyman, J. and Ross, D. (2007), *Every Thing Must Go* (Oxford University Press).

Langton, R. and Lewis, D. (1998), "Defining 'Intrinsic'," *Philosophy and Phenomenological Research* 58, 333–345.

LaPorte, J. (2000), "Rigidity and Kind," *Philosophical Studies* 97, 293–316.

(2004), *Natural Kinds and Conceptual Change* (Cambridge University Press).

Lewis, D. (1972/1999), "Psychophysical and Theoretical identifications," in *Papers in Metaphysics and Epistemology* (Cambridge University Press), 248–261.

(1983), "New Work for a Theory of Universals," *Australasian Journal of Philosophy* 61, 343–377.

(1986), "Against Structural Universals," *Australasian Journal of Philosophy* 64, 25–46.

List, C. and Menzies, P. (2007), "Non-Reductive Physicalism and the Limits of the Exclusion Principle," Working Paper, London School of Economics, Government Department, available at: http://eprints.lse.ac.uk/20118/.

List, C. and Menzies, P. (2009), "The Causal Autonomy of the Special Sciences," in G. Macdonald and C. Macdonald (eds.), *Emergence* (Oxford University Press), 108–128.

Locke, J. (1689/1824), *Essay Concerning Human Understanding* in *The Works of John Locke in Nine Volumes*, vols. I–II, 12th edn. (London: Rivington).

Longino, H. (1983), "Beyond 'Bad Science': Skeptical Reflections on the Value-Freedom of Scientific Inquiry," *Science, Technology & Human Values* 8, 7–17.

Lowe, E. J. (2004), "The Four-Category Ontology: Reply to Kistler," *Analysis* 64, 152–157.

(2006), *The Four-Category Ontology: A Metaphysical Foundation for Natural Science* (Oxford University Press).

Mallon, R. (2003), "Social Constructionism, Social Roles, and Stability," in Frederick Schmitt (ed.), *Socializing Metaphysics* (Lanham, MD: Rowman and Littlefield), 327–354.

Mandik, P. and Clark, A. (2002), "Selective Representing and World-Making," *Minds and Machines* 12, 383–395.

Marras, A. (2007), "Kim's Supervenience Argument and Nonreductive Physicalism," *Erkenntnis* 66, 305–327.

Mellor, D. H. (1977), "Natural Kinds," *British Journal for the Philosophy of Science* 28, 299–312.

Mill, J. S. (1843/1974), *A System of Logic* in *The Collected Works of John Stuart Mill*, vol. VII, ed. John M. Robson, Introduction by R. F. McRae (University of Toronto Press).

Miller, R. W. (2000), "Half-Naturalized Social Kinds," *Philosophy of Science*, 67 (Proceedings), S640–S652.

Millikan, R. (1984), *Language, Thought, and Other Biological Categories* (Cambridge University Press).

(1999), "Historical Kinds and the 'Special Sciences'," *Philosophical Studies* 95, 45–65.

(2000), *On Clear and Confused Ideas* (Cambridge University Press).

(2005), "Why Most Concepts Aren't Categories," in H. Cohen and C. Lefebvre (eds.), *Handbook of Categorization in Cognitive Science* (Amsterdam: Elsevier), 305–315.

Miron, L. F. (1996), *The Social Construction of Urban Schooling: Situating the Crisis* (Creskill, NJ: Hampton Press).

Mitchell, S. (2000), "Dimensions of Scientific Law," *Philosophy of Science* 67(2), 242–265.

Monck, W. H. S. (1887), "Mill's Doctrine of Natural Kinds," *Mind* 12, 637–640.

Moreira, D. and López-García, P. (2009), "Ten Reasons to Exclude Viruses from the Tree of Life," *Nature Reviews Microbiology* 7, 306–311.

Mukherjee, S. (2010), *The Emperor of All Maladies: A Biography of Cancer* (New York: Simon & Schuster).

Mumford, S. (2005), "Kinds, Essences, Powers," *Ratio* 18, 420–436.

Murphy, D. (2006), *Psychiatry in the Scientific Image* (Cambridge, MA: MIT Press).

Needham, P. (2000), "What Is Water?," *Analysis* 60, 13–21.

Newman, W. I. and Sagan, C. (1981), "Galactic Civilizations: Population Dynamics and Interstellar Diffusion," *Icarus* 46, 293–327.

Okasha, S. (2002), "Darwinian Metaphysics: Species and the Question of Essentialism," *Synthese* 131, 191–213.

Papineau, D. (1985), "Social Facts and Psychological Facts," in G. Currie and A. Musgrave (eds.), *Popper and the Human Sciences* (Dordrecht: Nijhoff), 57–71.

(2009), "Physicalism in the Human Sciences," in C. Mantzavinos (ed.), *Philosophy of the Social Sciences: Philosophical Theory and Scientific Practice* (Cambridge University Press), 103–123.

(2010), "Can Any Sciences Be Special?," in C. Macdonald and G. Macdonald (eds.), *Emergence in Mind* (Oxford University Press), 179–197.

Pettenger, M. E. (ed.) (2007), *The Social Construction of Climate Change: Power, Knowledge, Norms, Discourses* (Burlington, VT: Ashgate).

Pickering, A. (1984), *Constructing Quarks: A Sociological History of Particle Physics* (University of Chicago Press).

Putnam, H. (1973), "Meaning and Reference," *Journal of Philosophy* 70, 699–711.

(1975), *Mind, Language, and Reality: Philosophical Papers*, vol. II (Cambridge University Press).

Quine, W. V. (1948/1953), "On What There Is," in *From a Logical Point of View* (Cambridge, MA: Harvard University Press), 1–19.

(1969), "Natural Kinds," in *Ontological Relativity and Other Essays* (New York: Columbia University Press), 114–138.

Raatikainen, P. (2010), "Causation, Exclusion, and the Special Sciences," *Erkenntnis* 73(3), 349–363.

Rorty, R. (1999), *Philosophy and Social Hope* (London: Penguin).

Ross, D. (2000), "Rainforest Realism: A Dennettian Theory of Existence," in D. Ross, A. Brook, and D. Thompson (eds.), *Dennett's Philosophy* (Cambridge, MA: MIT Press), 147–168.

Ross, D. and Spurrett, D. (2004), "What to Say to a Skeptical Metaphysician: A Defense Manual for Cognitive and Behavioral Scientists," *Behavioral and Brain Sciences* 27, 603–647.

Ruddon, R. W. (2007), *Cancer Biology*, 4th edn. (Oxford University Press).

Ruphy, S. (2010), "Are Stellar Kinds Natural Kinds? A Challenging Newcomer in the Monism/Pluralism and Realism/Antirealism Debates," *Philosophy of Science* 77(5), 1109–1120.

Russell, B. (1913), "On the Notion of Cause," *Proceedings of the Aristotelian Society* 13, 1–26.

(1948), *Human Knowledge: Its Scope and Limits* (New York: Simon and Schuster).

Sagvolden, T., Johansen, E. B., Aase, H., and Russell, V. A. (2005), "A Dynamic Developmental Theory of Attention-Deficit/Hyperactivity Disorder (ADHD) Predominantly Hyperactive/Impulsive and Combined Subtypes," *Behavioral and Brain Sciences* 28, 397–419.

Salmon, N. (1981), *Reference and Essence* (Princeton University Press).

(2005), "Are General Terms Rigid?," *Linguistics and Philosophy* 28, 117–134.

Schaffer, J. (2003), "Is There a Fundamental Level?," *Noûs* 37, 498–517.

Schwartz, S. (1980), "Formal Semantics and Natural Kind Terms," *Philosophical Studies* 38, 189–198.

(2002), "Kinds, General Terms, and Rigidity," *Philosophical Studies* 109, 265–277.

Searle, J. (1995), *The Construction of Social Reality* (New York: Free Press).

Seidman, S. (2010), *The Social Construction of Sexuality*, 2nd edn. (New York: Norton).

Seppä, P. and Pamilo, P. (1995), "Gene Flow and Population Viscosity in *Myrmica* Ants," *Heredity* 74, 200–209.

Shapiro, L. (2000), "Multiple Realizations," *Journal of Philosophy* 97, 635–654.

(2004), *The Mind Incarnate* (Cambridge, MA: MIT Press).

Simon, H. A. (1962), "The Architecture of Complexity," *Proceedings of the American Philosophical Society* 106(6), 467–482.

Slater, M. H. (2009), "Macromolecular Pluralism," *Philosophy of Science* 76, 851–863.

(2010), "A Different Kind of Property Cluster Kind," paper delivered at the Biennial Meeting of the Philosophy of Science Association, Montreal, November 2010.

Smolin, L. (2006), *The Trouble with Physics* (New York: Houghton Mifflin).

Soames, S. (2002), *Beyond Rigidity: The Unfinished Semantic Agenda of Naming and Necessity* (Oxford University Press).

Sober, E. (1980), "Evolution, Population Thinking, and Essentialism," *Philosophy of Science* 47, 350–383.

(1999), "The Multiple Realizability Argument Against Reductionism", *Philosophy of Science* 66, 542–564.

Steinberg, P. E. (2001), *The Social Construction of the Ocean* (Cambridge University Press).

Sterelny, K. (1983), "Natural Kind Terms," *Pacific Philosophical Quarterly* 35, 110–125.

Stich, S. (1990), *The Fragmentation of Reason: Preface to a Pragmatic Theory of Cognitive Evaluation* (Cambridge, MA: MIT Press).

Thomasson, A. (2003a), "Foundations for a Social Ontology," *Protosociology* 18–19, 269–290.

(2003b), "Realism and Human Kinds," *Philosophy and Phenomenological Research* 67, 580–609.

Towry, M. H. (1887), "On the Doctrine of Natural Kinds," *Mind* 12, 434–438.

Vallentyne, P. (1997), "Intrinsic Properties Defined," *Philosophical Studies* 88, 209–219.

Van Valen, L. (1976/1992), "Ecological Species, Multispecies, and Oaks," in Ereshefsky (ed.), *Units of Evolution*, 69–78.

Venn, J. (1876), *The Logic of Chance*, 2nd edn. (London: Macmillan).

(1889), *The Principles of Empirical or Inductive Logic* (London: Macmillan).

Weatherson, B. (2006), "Intrinsic vs. Extrinsic Properties," in E. N. Zalta (ed.), *Stanford Encyclopedia of Philosophy*, http://plato.stanford.edu/entries/intrinsic-extrinsic/

Weisberg, M. (2006), "Water Is Not H$_2$O," *Boston Studies in the Philosophy of Science* 242, sect. VII, 337–345.

Weiskopf, D. A. (2011), "The Functional Unity of Special Science Kinds," *British Journal for the Philosophy of Science* 62, 233–258.

Whewell, W. (1847), *The Philosophy of the Inductive Sciences*, vol. II (London: Parker).

Wilkerson, T. E. (1988), "Natural Kinds," *Philosophy* 63, 19–42.

(1993), "Species, Essences, and the Names of Natural Kinds," *Philosophical Quarterly* 43: 1–19.

(1998), "Recent Work on Natural Kinds," *Philosophical Books* 39, 225–233.

Williams, N. E. (2011), "Arthritis and Nature's Joints," in J. K. Campbell, M. O'Rourke, and M. H. Slater (eds.), *Carving Nature at Its Joints* (Cambridge, MA: MIT Press), 199–230.

Wilson, R. A. (ed.) (1999), *Species: New Interdisciplinary Essays* (Cambridge, MA: MIT Press).

Wilson, R. A., Barker, M. J., and Brigandt, I. (2009), "When Traditional Essentialism Fails: Biological Natural Kinds," *Philosophical Topics* 35, 189–215.

Wimsatt, W. C. (1994/2007), "The Ontology of Complex Systems: Levels of Organization, Perspectives, and Causal Thickets," in *Re-Engineering Philosophy for Limited Beings* (Cambridge, MA: Harvard University Press), 193–240.

Woodward, J. (2003), *Making Things Happen: A Theory of Causal Explanation* (Oxford University Press).

Yablo, S. (1992), "Mental Causation," *Philosophical Review* 101(2), 245–280.

Zemach, E. M. (1976), "Putnam's Theory on the Reference of Substance Terms," *Journal of Philosophy* 73, 116–127.

Zhang, H. M. (2003), "Driver Memory, Traffic Viscosity and a Viscous Vehicular Traffic Flow Model," *Transportation Research: Part B* 37, 27–41.

Index